The Economics *of* Groundwater Remediation *and* Protection

Integrative Studies in Water Management and Land Development

Series Editor
Robert L. France

Published Titles

Handbook of Water Sensitive Planning and Design
Edited by Robert L. France

**Boreal Shield Watersheds: Lake Trout Ecosystems
in a Changing Environment**
Edited by J.M. Gunn, R.J. Steedman, and R.A. Ryder

**Forests at the Wildland–Urban Interface:
Conservation and Management**
Edited by Susan W. Vince, Mary L. Duryea, Edward A. Macie,
and L. Annie Hermansen

The Economics of Groundwater Remediation and Protection
Paul E. Hardisty and Ece Özdemiroğlu

Forthcoming Titles

Restoration of Boreal and Temperate Forests
Edited by John A. Stanturf

Stormwater Management for Low Impact Development
Edited by Lawrence Coffman

Porous Pavements
Bruce K. Ferguson

The Economics of
Groundwater Remediation
and Protection

Paul E. Hardisty and Ece Özdemiroğlu

CRC PRESS

Boca Raton London New York Washington, D.C.

Library of Congress Cataloging-in-Publication Data

Hardisty, Paul E.
  The economics of groundwater remediation and protection / Paul E.
Hardisty and Ece Özdemiroğlu.
    p. cm. — (Integrative studies in water management and land development)
    Includes bibliographical references and index.
    ISBN 1-56670-643-2 (alk. paper)
    1. Groundwater—Purification—Economic aspects. I. Özdemiroğlu, Ece. II. Title. III.
Integrative studies in water management and land development

TD426.H37 2005

363.739'4—dc22   2004056685

Visit the CRC Press Web site at www.crcpress.com

© 2005 by CRC Press

No claim to original U.S. Government works
International Standard Book Number 1-56670-643-2
Library of Congress Card Number 2004056685
Printed in the United States of America 1 2 3 4 5 6 7 8 9 0
Printed on acid-free paper

Dedications

To Heidi, Zachary, and Declan

Paul Hardisty

To my aunty Oya

Ece Özdemiroğlu

Series statement: Integrative studies in water management and land development

Ecological issues and environmental problems have become exceedingly complex. Today, it is hubris to suppose that any single discipline can provide all the solutions for protecting and restoring ecological integrity. We have entered an age where professional humility is the only operational means for approaching environmental understanding and prediction. As a result, socially acceptable and sustainable solutions must be both imaginative and integrative in scope; in other words, garnered through combining insights gleaned from various specialized disciplines, expressed and examined together.

The purpose of the CRC Press series Integrative Studies in Water Management and Land Development is to produce a set of books that transcends the disciplines of science and engineering alone. Instead, these efforts will be truly integrative in their incorporation of additional elements from landscape architecture, land-use planning, economics, education, environmental management, history, and art. The emphasis of the series will be on the breadth of study approach coupled with depth of intellectual vigor required for the investigations undertaken.

Robert L. France
Series Editor
Integrative Studies in Water Management
and Land Development
Adjunct Associate Professor of Landscape Ecology
Science Director of the Center for
Technology and Environment
Harvard University
Principal, W.D.N.R.G. Limnetics
Founder, Green Frigate Books

Foreword by series editor: out of sight but not out of mind

Driving northward from Boston along the Interstate towards the mountain greenery of New Hampshire and Vermont, one passes nearby a site of worldwide historic significance with respect to groundwater pollution and economic repercussions. There, just off the side of the highway, admix the cluster of factories easily recognized by their dense array of pipes and vents jutting into the air, is the location of events made infamous in the best-selling book by Jonathan Harr, *A Civil Action* (later filmed with John Travolta and Robert Duvall). If one pulls off the highway and drives or walks around the town of Woburn, there is nothing obvious to signify the pollutant-related deaths that occurred there several decades ago, nor are there any environmental scars upon the surface of the land to indicate damage done by the various factories and tanneries. This is because all the damage took place underground, hidden out of sight, due to the contamination of groundwater upon which residents depended for drinking. Indeed, as the case made its torturous way through the courts (described with page-turning detail in Harr's book), the hydrogeologists were busy inserting thousands of sampling devices into the ground (the greatest number of such anywhere on the Earth) in order to try to map the pollution plume such that they could either prove or disprove a linkage between industrial discharge and certain drinking wells that had already identified by the epidemiologists as being sources for the leukemia ravaging the community. Though eventually the case would lead to the largest environmental settlement in United States history, all sides would have greatly benefited from a careful reading of the present book had it have been available at that time.

As the fifth book and first authored (i.e., nonedited) title in the series by CRC Press – Integrative Studies in Water Management and Land Development – the present volume by Paul Hardisty and Ece Özdemiroğlu brings much needed attention to the plight and remediation of one of the world's most important but overlooked sources of water. The authors lay out a logical framework of background knowledge that seamlessly fuses the disciplines of hydrogeology and economics. The illustrative case studies are presented in thorough detail to enable readers to understand not only the product of the remedial measures being undertaken, but also, just as importantly, the process toward achieving those ends.

The basic idea underlying this book is that economic analysis can be used to aid decision making for environmental protection and restoration. Although that may be the overt intended utility, the other important value put forward is one of education in terms of fostering awareness as the means toward facilitating management, a subject I have explored in my book *Facilitating Watershed Management: Fostering*

Awareness and Stewardship. The challenge, as so clearly enunciated by the authors of the present volume, is that the resource they are endeavoring to protect and manage lies out of sight and, therefore, often out of mind. This book does a wonderful service in bringing groundwater up from the depths of its "forgetfulness" (*sensu* the philosopher Ivan Illich as cited in the Introduction) to the surface of our collective consciousness. Too often we as a society have adopted a cavalier attitude towards contaminating groundwater; i.e., sweeping, as it were, the dirt of our industrial lives under the imagined protective rug of the ground. What this book does so well is to educate us about the misfortunes and mitigations of such a disposal strategy. And, as I have argued elsewhere (*Deep Immersion: The Experience of Water*), it is necessary to first recognize something in order to value it, just as it is necessary to value something before one will protect it. In this light, the present volume does an admirable job of helping us with that first stage of recognition of groundwater just as much as it succeeds in outlining a program toward protecting the resource. And for this we should be indebted to the authors of this important volume.

Robert L. France
Harvard University

Preface

Of all the fresh water on Earth, two-thirds is locked up as snow and ice at the polar caps. Of the remaining amount, half exists below ground, seeping slowly through the tiny pores and fractures in the rock that makes up the Earth's crust. This is groundwater, and it exists within layers of porous soil and rock that scientists call aquifers. Many countries in the world, including the United States and the United Kingdom, depend heavily on groundwater as a source of water supply. All of the bottled and delivered spring water many of us drink is from deep underground, pumped to surface through wells or collected at spring discharges in the mountains. In more arid parts of the world, groundwater often forms the only dependable and accessible source of fresh water.

Groundwater is particularly important because of its dependability and purity. Despite this, it is not always well protected from the pressures human activities impose on it. Once threatened by overabstraction or contamination, groundwater is very difficult to remediate, not only because it is difficult to work underground but also because natural processes mean that it may be too late by the time we realize that the aquifer is damaged.

The multiple uses of groundwater mean that remediation raises a wide range of environmental, social, and economic issues, affecting a large number of stakeholders. This is also reflected in the research and policy interest in groundwater by scientists, environmental engineers, economists, lawyers, polluters, regulators, and nongovernmental organizations (NGOs), to name but a few. As would be expected, different groups champion different sides of the argument. But more importantly (and in our view somewhat counterproductively), different groups use such different terminologies that they can almost be classified as distinct languages. Specifically related to the topic of this book, the most obvious example of the existence of at least two different languages in this discourse is the use of the word *economic*. To scientists and engineers, an economic remediation option means the cheapest way to remediate contamination. Economists, on the other hand, undertake an economic analysis to find out which remediation option generates the maximum benefit in terms of the avoided damage from contamination. It is not difficult to imagine that the two options (the cheapest and the one that generates the maximum benefits or avoids the most damage) may not always be — in fact, rarely are — the same. Benefits here refer not only to those that accrue to parties responsible for contamination and remediation (the problem holder) but also to the rest of the society.

Although scientists and engineers must use disparate units of measure to identify the different impacts of contamination (e.g., through risk assessment) and compare these to the cost of remediation, economists try to avoid the old adage that you cannot compare apples and oranges, expressing as many of the impacts as possible in one common unit: money. Money is a unit that everyone may value differently but, nonetheless, understands and is familiar with. This book aims to use a common language, one that adheres to economists' definitions of *economic*, *benefits*, and *costs*. This discussion is not limited to the use of money as a measuring rod, but includes the conceptual approaches to decision making.

This book also shows how different stakeholders could use this common language for a more encompassing decision-making framework. The problem holders could benefit from including the effect of contamination on others in their remedial decision-making process. Regulators could set remediation objectives that are not excessively onerous on the problem holder, and generate sufficient benefits for the rest of the society. The public and environmental groups could get their demands across more forcefully by speaking the language of business. All of these different groups (and others), following a common framework if not yet speaking fluently a common language, may in turn provide additional incentive to all sectors to consider seriously and act on their legitimate environmental responsibilities, in the knowledge that at the end of the day, environmental pollution and resource damage have an economic impact on all of us.

We, a hydrogeologist and an economist, started to use this common language in our work for both private- and public-sector clients about six years ago. Having gathered sufficient evidence since then that the common language and framework worked well for us and our clients, we wanted to share our experiences with others. Although this book focuses on some sectors more than others, we believe that the framework applies equally well to all. We hope it will help you as much as it does us.

PEH, Nicosia
EO, London

Acknowledgments

The authors would like to thank Dr. Steve Wallace of Secondsite Property Plc for his steadfast support, Tom Parker, Simon Firth, and everyone else at Komex, all at eftec who contributed in one way or another to this book, Jonathan Smith of the U.K. Environment Agency, who was instumental in setting up and guiding the initial research contract that led to this book, and Dave Drabble.

Authors

Paul Hardisty, Ph.D., is managing director for Europe and International, Komex Environmental Ltd., a global environmental consulting firm. He obtained a B.Sc. in geological engineering from the University of British Columbia; an M.Sc. in engineering hydrology from the Imperial College of Science and Technology, London; and a Ph.D. in environmental engineering from the University of London. He has more than 17 years of experience in the environmental field, specializing in groundwater resources development, protection, and remediation. Recently, Hardisty completed a major study for the United Kingdom. Environment Agency on the economics of groundwater remediation. He has advised the Canadian and Ontario governments and major industrial corporations in the U.K. and North America on the subject. He has designed and led over 200 groundwater-related projects throughout Europe, North America, the Middle East, and South America. Hardisty is a visiting lecturer in contaminant hydrology at Imperial College, London, and a research associate at Trinity College, University of Dublin. He is the author of numerous papers and articles on groundwater protection, remediation, and economics. He is a licensed professional engineer and a member of the Association of Ground Water Scientists and Engineers, a division of NGWA; the International Association of Hydrogeologists; and the International Union of Geophysicists.

Ece Özdemiroğlu, M.Sc., is the founding director of Economics for the Environment Consultancy (eftec), the United Kingdom's leading consultancy firm specializing in environmental economics. She obtained a B.A. in economics from the University of Istanbul and an M.Sc. in environmental economics and natural resource management from University College London. She has 12 years of experience in environmental economics, specializing in estimating economic costs and benefits of environmental impacts and economic appraisal of policies and projects. Ece has undertaken over 20 cost–benefit analyses in a variety of sectors, such as air quality, water supply, nature conservation, and groundwater remediation in Europe, North America, and the Middle East for government agencies, private companies, and international financial institutions. She is the coauthor or coeditor of 10 books and numerous articles and conference papers. She is a member of the U.K. Network of Environmental Economists.

Table of Contents

PART II
Applying Economics to Groundwater 89

PART V
Summary and Conclusions 309

Part I

Introduction

1 The Case for Rational Environmental Decisions

1.1 THE CHALLENGE OF ENVIRONMENTAL PROTECTION

As the world's population grows, the demand for resources rises and, inevitably, so does the impact of our activities on the earth's natural environment. The use of resources is necessary to creating well-being for the people who share the planet. Each individual's struggle to improve his or her standard of living and that of any progeny is fundamental to the human experience. But as our numbers grow and we become more and more successful at harnessing the earth's bounty, the health of the natural environment that sustains us all becomes increasingly precarious. Contrary to the widely held beliefs of previous centuries, the earth's resources are not limitless. Overexploitation, poor harvesting and management practices, and waste generation all threaten our resource base. Every facet of modern life — from mechanized agricultural production, to the extraction and refining of petroleum, to the manufacture of our most basic consumer goods — creates wastes and by-products that must be disposed of and managed. Inevitably, some of these wastes find their way back into the natural environment, sometimes with negative results.

The challenge of this and future generations will be to increase living standards for all the world's people, most notably its poorest, while sustaining and protecting the natural ecosystems upon which our prosperity depends. Without clean water, breathable air, stable climates, fertile soil, and thriving biodiversity, it is doubtful whether living standards can be maintained at their current levels, let alone improved.

1.2 WATER

At dawn in the small village of Wawase in the Ashanti province of Ghana, West Africa, young women and girls start the first of their daily trips down the road to fetch water. The trip back is difficult, uphill and laden with jars and cans of the precious life-giving fluid, heavy loads balanced elegantly on the carriers' heads. In the last decade, courtesy of loans from the World Bank, Wawase has been provided with several new wells, tapping the clean but slightly acidic groundwater from the rock below (Figure 1.1). For the women, the journey to the hand-pump is now much shorter than it was when they used the waterhole, a murky depression fed by springs and surface water, a breeding ground for diseases such as bilharzia, guinea worm, and yellow fever. Wawase is typical of small villages across Africa and in many other developing regions. Life has improved because of the wells. But fetching water is still hard work, and water must be used sparingly. In fact, people in this part of Africa use less than 15 liters

FIGURE 1.1 Groundwater is often the only reliable source of clean water for many rural communities, especially in less developed countries.

each of water per day for domestic purposes such as washing, cooking, and cleaning, compared with over 60 liters per day (and rising quickly) for the average Cypriot, and a colossal 600 liters per day for the average American.

And yet, the people of Wawase are among the lucky ones. The United Nations, in its 2003 report "Water for People, Water for Life," estimates that over a billion people in developing countries have inadequate access to water, and 2.4 billion lack basic sanitation. It is a deadly combination. Inadequate water supply and poor or nonexistent sanitation measures, including treatment of wastewater and sewage, provide the conditions where waterborne diseases flourish. Two billion people, mostly children, suffer from acute gastrointestinal diseases caused by dirty water every year, and many die. Two hundred million a year are inflicted with schistosomiasis, and 1.5 billion a year with intestinal helminths (worms of various kinds). At the Johannesburg Earth Summit in 2002, the world's nations pledged to reduce these numbers by half by 2015. Worthy goals, but ones that the World Bank estimates will cost over US$600 billion for infrastructure alone (less than the world spends on arms in six months).

Of all the water on the planet, only 3% is fresh, and of that the majority is locked away as snow and ice at the poles. Less than a third of the fresh water on earth is actually available to support the ecosphere, flowing in lakes and rivers, falling as rain and filtering slowly through underground rocks as groundwater. The sun powers a continuous hydrologic cycle, evaporating fresh water from the seas and driving it back to earth as rain.

Fresh water is a renewable resource, but a finite one and unevenly distributed. Some parts of the world, like Canada and Finland, are blessed with more water than they know what to do with. Others, like sub-Saharan Africa and the Middle East, experience chronic shortages (Figure 1.2). In ancient times, civilizations were born and flourished in places where water was plentiful and available. The city of Sana'a in Yemen was founded in a wide valley surrounded by mountains of porous and permeable volcanic rocks, where perennial springs bubbled up pure and sweet from the ground (but no longer — overpumping has dried up the springs and the shallow aquifer is now badly contaminated with sewage). But as populations expanded and the needs of agriculture grew, water had to be harvested and moved to where it was needed. From Roman aqueducts, to Victorian distribution networks, to the major dam and interbasin transfer schemes of recent times, capturing and moving water has been a human preoccupation on the grandest of scales.

The main problem with water, according to the *Economist,*[1] is not scarcity, but "man's extravagantly wasteful misuse of it." In their recent survey of the world's water, the *Economist* argues that water has been ill governed and hugely underpriced. Cheap, or indeed even free water, encourages waste and misallocation and ignores the huge costs of the dams, reservoirs, pipelines, and pumping systems needed to deliver it. It also leads to using water for the wrong things, in the wrong places. Irrigation of water-intensive crops, using inefficient methods, is one of the worst offences. The Mediterranean region is a case in point. The area is in the midst of a protracted drought that began almost 30 years ago. And although it is raining less, a lot more water is used. Again, echoing the trends in many other regions, agriculture accounts for a large proportion of water withdrawn, much of it for unlicensed small, private farms and the growing of thirsty crops such as bananas. Perhaps the worst example of water misuse is the cultivation of wheat in Saudi Arabia, using water from oil-fired desalination plants. In that climate, it takes 1000 tons of water to

FIGURE 1.2 Groundwater supply for rural communities, Hadramout, Yemen.

produce one ton of wheat, making the real cost of Saudi wheat about 100 times the world price.

But water is heavy, and transporting it is energy intensive and expensive. The ladies of Wawase know this all too well. Faced with a difficult and tiring journey several times each day, these women carry only what they need and waste not a drop. They are exercising an economic decision, which we all would if the true price (the energy and effort of carrying a heavy load several kilometers) had to be paid. There are many examples of effective and fair water-pricing policies. Australia and South Africa, two of the driest countries on earth, have developed water pricing and allocation policies that the *Economist*[1] considers among the most progressive. Both have shunned privatization of water delivery services, along the lines of the U.K. model, in favor of strong state control of water issues through effective water laws. Australia is the only country in world with 100% water metering, allowing them to charge by volume, encouraging efficient use. South Africa has opted to deliver the first 25 liters completely free, to ease the burden of the poorest families, whose women would have to walk several miles rather than pay even the smallest charge.

We cannot forget that water is also vital to all other forms of life on earth. And unlike us, our fellow creatures cannot shape their world — they depend entirely on finely balanced ecosystems, to which they have adapted over millennia. Our attempts to harvest water and move it to where we need or want it most have vastly reshaped the natural hydrology and ecology of huge parts of the planet. The Colorado River in the U.S., one of the largest in the world, has been dammed up, diverted, and its water allocated and siphoned off to the extent that not a drop now flows into the Gulf of Mexico. The Yellow River in China and the Ganges in India suffer the same fate. The Aral Sea in Central Asia has been virtually destroyed since the 1950s, when Soviet engineers started to divert the two rivers that fed the Aral. The plan was to use the water to grow cotton, a notoriously "thirsty" crop, in the desertlike plains of Khazakstan. What resulted was one of the worst environmental disasters ever. The Aral Sea has shrunk to nearly half its area, losing over 70% of its volume. Its once fresh water has become so salty that all the fish have died, and the dried-out seabed has turned into a dustbowl.[2]

Besides being fundamental to all life, water is also a unique molecule. It is one of the few compounds that actually expands when it freezes, causing ice to float. If it were not so and water acted like most other compounds, the ice caps would sink, and the earth would actually be covered by a single, vast ocean, dotted with a few lonely islands. Water is also a strong solvent. The polar arrangement of its twin hydrogen atoms, attached in a V shape to the lone oxygen atom, gives water the ability to dissolve a huge range of compounds, from sugar to limestone. When salts are dissolved into water, it is able to conduct electricity and thus transmit the electrochemical signals on which all life depends. Human beings are 98% water, and without a steady supply of the stuff, we would be dead in under a week. Its properties as a solvent also make water ideal for washing. Its powers of cleansing and purification have been revered by mankind since earliest times. Many of our religious ceremonies have evolved to include water, from ritual ablutions to blessings and baptisms.

But ironically perhaps, water has also become our favored medium for moving waste. In his essay on water, Illich describes the transition of water from sacred to defiled.[3] Where once water had an almost spiritual power, giving life and sustaining communities, it has now become a carrier for every imaginable form of refuse, from sewage to toxic chemicals. This most precious of resources is under threat worldwide. Scarce and valuable fresh water is being polluted at an alarming rate, putting even greater stress on the supplies that remain. Protecting our water supplies, rivers, lakes, and underground aquifers from pollution has become a major issue worldwide.

Common sense and economic theory tell us that when supply is limited, the value of what remains increases. More and more, this is becoming the case with the natural environment. The science of economics is now being used as a way of putting environmental protection into context with the legitimate needs of human beings. This book is about using economics to make better decisions about the protection and restoration of one small part of our environment: the water that is found beneath the earth's surface — groundwater.

1.3 GROUNDWATER IN CONTEXT

Groundwater is one of our most precious resources and one of our most fragile. Reclaiming polluted aquifers can be expensive, technically difficult, and time consuming. Deciding if and when to remediate, and to what degree, can be regarded in the context of alternative environmentally and socially beneficial actions. What else can be done with the money required to restore an aquifer? Could we purchase and preserve several acres of rainforest? Could we preserve or restore a few hundred acres of natural habitat? Could we equally fund environmental awareness programs? Which of these options would provide the greatest benefit to society? And then, what are the commercial realities facing those who are called upon to pay for the restoration of polluted groundwater? Under the "polluter pays" principle, increasingly adopted as the fundamental ethical precept for remediation policy, the responsibility for planning, funding, and executing remediation must be borne by the polluter (Figure 1.3). This could be a state or federal government, a municipality, or a private-sector enterprise. And in the background, ever present and increasingly vocal and powerful, are the public, the neighbors, the inhabitants of the planet, demanding rightly that their interests be served and the planet's dwindling resources protected for their future and the future of their children and grandchildren. Combining and prioritizing these diverse interests into a decision-making process, using a common unit of value, is essential if equitable, practical, and rational economic decisions are to be made.

Despite the vast sums spent on managing contaminated land and groundwater pollution in the past 15 years, a detailed review of the literature shows that relatively little research has been conducted into applying cost–benefit techniques to these problems.[4] The available literature has been produced by economists or by technical (scientific and engineering) experts but shares little common ground. Not unexpectedly, the economic literature deals mainly with valuation of groundwater and land, and with the external economic benefits of groundwater protection. Some work deals directly with remediation. The technical–scientific literature focuses on the applica-

FIGURE 1.3 Excavation of contaminated soil as part of a major remediation project in the U.K.

tion of specific techniques and technologies and deals almost entirely with remedial costs, cost comparisons, and cost effectiveness. The wider benefits of remediation are rarely discussed. Much of this work is of primary interest to problem holders, but even so, very little is available that discusses the private benefits of remediation.[4]

Especially when one is considering highly mobile contaminants, such as the gasoline additive MtBE (methyl-tert butyl ether), or expensive remediation of hydrocarbon NAPLs (nonaqueous phase liquids) in deep fractured bedrock, the need for a complete analysis of the economics of site contamination and a rational, objective analysis of the full costs and benefits of remediation becomes apparent.

1.4 SUSTAINABILITY, OPTIMALITY, AND INTERVENTION

When considering the economics of groundwater remediation and protection, the concepts of *sustainability*, *optimality*, and *intervention* are particularly relevant and are introduced here. In common environmental parlance, the word sustainability has come to describe the notion of using a resource responsibly, such that future generations may enjoy equal access. In economic terms, the word sustainability describes much the same concept, but is more strictly defined as a state where per-capita welfare is nondecreasing (increasing or at least staying constant) over time. Therefore, invoking sustainability, only those projects that increase welfare should go ahead. Sustainable development implies that there is no net reduction of the resource base where this includes natural, man-made, human, and even social capital.

Groundwater is usually considered to be a renewable resource, and so its sustainable use is preferred. Contamination of groundwater may have the effect of destroying the resource or eliminating it from use, and so can be seen as an unsustainable activity.

Economic theory distinguishes between project proponent and the rest of society. In a typical example of groundwater contamination, the proponent would be the problem holder (the group responsible for the contamination and thus the clean-up). Economists call the proponent's view *private*. The private optimum is where the net present value of the proponent's welfare is maximized, and the social optimum is where society's net present value of welfare is maximized (the social includes the private). The concept of present value accounts for discounting of future welfare gains, which are considered less important the further in the future they occur. Discounting is discussed in more detail in Chapter 5.

In a private project analysis, the effects of the project on third parties are not taken into account. These uncompensated effects on third parties are known as *externalities*. In the case of groundwater contamination, the polluters' optimum may not include the loss of welfare experienced by society as a whole, including users of impacted groundwater and surface water. This is one of the fundamental differences between a purely financial analysis of groundwater remediation, of which the literature abounds, and a truly complete economic analysis. In an economic analysis, the benefits to *all stakeholders* are considered.

Figure 1.4 shows a schematic of the relationships between sustainable and unsustainable policies or practices, private and social optima, and the results of various types of intervention. Pearce and Warford explain that, typically, "private and social optima diverge: the most desirable rate at which to deplete resources from the standpoint of their owner is unlikely to be the best rate for society as a whole."[5] For example, in Case 1 (Figure 1.4) while private optimum is sustainable, government intervention (in the form of economic policy or setting technical standards) can further improve sustainability by achieving the social optimum. Case 2 shows a position where undertaking private (financial) analysis alone would lead to an unsustainable outcome, which could be changed to sustainable social optimum by intervention. However, achieving the social optimum does not necessarily provide sustainability (see Case 3 in Figure 1.4 where both private and social optima are not sustainable). With groundwater remediation, economic analysis can be used to identify where the private and social optima diverge and thus rationally consider, in each case, what would be best for all stakeholders, using a common unit of measure that everyone understands — money.

1.5 ORGANIZATION OF THE BOOK

This book seeks to provide a comprehensive review of the state of the art of applying economic analysis to issues of groundwater remediation and protection. The book consists of four main parts, each with sequentially numbered chapters.

In Part I, groundwater is introduced as an issue of world importance, and basic background is given in economic theory and basic hydrogeology and remediation science. This background is intended to help nonhydrogeologists understand and

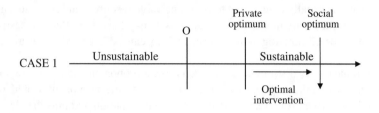

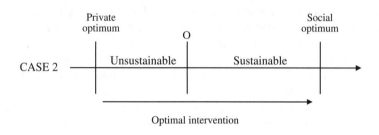

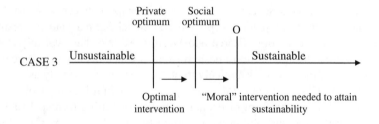

Rate of growth of welfare

FIGURE 1.4 Sustainability, optimality, and government intervention.

follow some of the technical issues discussed in later sections, and the economic background is intended to allow noneconomists to come to grips with important concepts key to conducting cost–benefit analyses.

Part II provides a closer look at applying economic concepts to groundwater problems and a detailed review of relevant available literature from both the economic and the hydrological disciplines. Costs and benefits of groundwater remediation and protection are introduced and examples provided.

Part III is the core of the volume. It lays out a comprehensive, step-by-step process for applying cost–benefit analysis as an aid to groundwater remediation decision making.

In Part IV, a series of case histories is provided, illustrating the application of cost-benefit analysis to problems of groundwater remediation. Examples of varying degrees of complexity and different levels of analysis are provided to illustrate the methods' range of applicability.

To the authors' knowledge, this is the first textbook to look specifically at this interesting and complex issue. It is first simply because this is a very new area of study, and correspondingly much remains to be done. Many aspects of this emerging subject remain to be studied, researched, and developed. Although the book is by no means complete in its consideration of the topic, it does attempt to provide an introduction to the subject, offer some guidance on how issues of groundwater remediation can be considered in an economic context, and put forward a few examples of how such analyses may be done and the results that can be achieved.

REFERENCES

1. *The Economist*, Survey: Water — Priceless, Print edition, London, Jul. 17, 2003.
2. De Villiers, M., *Water: The Fate of Our Most Precious Resource*. Mariner Books, New York, 2001.
3. Illich, I., *H₂O and the Waters of Forgetfulness*. Harper & Row, New York, 2001.
4. Komex and eftec, *Costs and Benefits Associated with Remediation of Contaminated Groundwater: A Review of the Issues*, EA R&D Technical Report P278, Bristol, 1999.
5. Pearce, D.W. and Warford, J.J., *World Without End*, World Bank, Oxford University Press, Oxford, 1993.

2 Contaminated Groundwater — A Global Issue

2.1 A NONTECHNICAL INTRODUCTION TO GROUNDWATER CONTAMINATION

Of all the fresh water on earth, two-thirds is locked up as snow and ice at the polar caps. Of the remaining amount, half exists below ground, seeping slowly through the tiny pores and fractures in the rock that makes up the earth's crust. This is groundwater, and it exists within layers of porous soil and rock that scientists call *aquifers*. Many countries in the world, including the U.S. and the U.K., depend heavily on groundwater as a source of water supply. In more arid parts of the world, groundwater often forms the only dependable and accessible source of fresh water (Figure 2.1A and Figure 2.1B).

Groundwater is particularly important because of its dependability and purity. The bottled spring water many of us drink begins its life as rainfall and snow, falling high in the mountains. Much runs off into streams or is evaporated away, but a significant proportion, as much as 20%, finds its way into the ground, percolating slowly over months and years through pores and fractures deep into the rock. Water moves slowly through the permeable geological formations, usually flowing only a few meters each year. As the water percolates through complex networks of pores and fissures, suspended particles and impurities are filtered out and harmful bacteria and other organic compounds destroyed, so that what emerges from our mountain springs is pure and safe to drink. Natural spring water has caught on as a commercial commodity. A liter of bottled spring water now costs more than a liter of gasoline in many places in the world. Consumers are paying a healthy premium because they believe, and the purveyors claim, that bottled spring groundwater is pure and free of toxins and other harmful chemicals.

But because groundwater is hidden underground, out of sight, it is often disregarded, misunderstood, or simply forgotten. Some consider it limitless, a never ending supply to be exploited simply by switching on a pump. Others see it as immune from damage, protected by thick layers of rock and earth. Unfortunately, groundwater is neither inexhaustible nor invulnerable. Groundwater can be polluted by a wide variety of human activities (Figure 2.2). Gasoline and other motor fuels, such as diesel, are a case in point. When fuel leaks out from an underground storage tank or buried pipeline and enters the ground, it behaves in a very characteristic way, governed by the fundamental laws of physics and chemistry. First, hydrocarbon

FIGURE 2.1A Drilling for deep groundwater in Yemen. (Photo courtesy of Komex.)

FIGURE 2.1B A water well in the Algerian Sahara Desert. Here, groundwater provides the only reliable source of water. (Photo courtesy of Komex.)

liquids move downward through the unsaturated soil, pulled by gravity. Hydrocarbon slowly fills up the pores in the soil, moving steadily downward. Like water being poured slowly onto a sponge, the pores of the sponge will gradually fill up, and

FIGURE 2.2 Natural gas processing plant. Petroleum refining and distribution facilities can impact groundwater through releases of hydrocarbons, produced water (brines), and other wastes and by-products. The petroleum industry is active in improving its environmental management through a number of organizations, such as IPIECA (International Petroleum Industry Environmental Conservation Association), the API (American Petroleum Institute), and CAPP (Canadian Association of Petroleum Producers).

eventually it will become saturated and the water will start to drip from the bottom. In the same way, if the spill volume is large enough, the petrol will eventually reach the groundwater table, the depth at which the ground becomes saturated. Being lighter than water, petrol will float on the groundwater table, accumulating in thickness, and then will start to move laterally in the direction of groundwater flow. Major spills can result in layers of fuel several meters thick "floating" underground on the water table, migrating slowly. Standing on the surface above, everything looks normal, but a few meters underground lies a potential environmental worry.

Although the liquid hydrocarbon itself typically moves much more slowly through the ground than water, due to viscous and capillary forces, most petrol typically contains as much as 15% benzene, which is partially water soluble and is known to cause cancer in humans. As water flows past the hydrocarbon, it dissolves and carries away compounds like benzene, xylene, and more complex and difficult-to-pronounce chemicals. Even at concentrations as low as a few parts per billion, these compounds will spoil the taste of water, rendering it unfit for human or even animal consumption. The World Health Organization limit for benzene in drinking water is only five parts per billion. This is equivalent to about one teacup of benzene mixed into ten Olympic-sized swimming pools. A little groundwater contamination can go a long way, literally and figuratively.

What makes this example of a petrol spill so relevant is that motor fuels are ubiquitous. Every town and city in every country in the world has its petrol filling

FIGURE 2.3   Petrol service station underground tank replacement.

stations (Figure 2.3). When you pull your car up to the pump and squeeze the handle on the nozzle, you are pumping fuel up from large tanks buried underground beneath the station forecourt. According to studies by the American Petroleum Institute and other industry organizations, these storage tanks have an unfortunate habit of leaking.[1] Leaking tanks were such a common problem that the U.S. government initiated in the mid-1980s a series of remediation programs for leaking underground storage tanks. Hundreds of millions of dollars have been spent across America during the last two decades, investigating and cleaning up petrol leaking into groundwater. However, the American example is also one of misplaced effort. Much of the clean-up work undertaken over the last 20 years was probably unnecessary or excessive, the result of overstrict regulations and inflexible enforcement. The overall economics of the effort were not fully considered at the outset. The economic benefit of the program as a whole likely did not justify the costs incurred.

In many parts of the world, agriculture is a major contributor to groundwater pollution, as well as the major groundwater user. In Europe, contamination of aquifers by nitrogen compounds from fertilizers is widespread and chronic and has been identified as a major challenge by the European Environment Agency. Increasingly, persistent and often toxic herbicides and pesticides are showing up in groundwater supplies, the result of careless and uncontrolled use by farmers and households. A wide variety of industries — from mining to steel making to computer-chip manufacturing to wood preserving — use and dispose of chemicals that can, and in some cases do, contaminate groundwater (Figure 2.4 and Figure 2.5).

Perhaps the biggest concern with underground contamination is that we cannot see it when it is happening, and we do not know where it is going until it gets there.

FIGURE 2.4 Uncontrolled waste disposal site in arctic Canada. Unlicensed waste tipping has occurred in almost every part of the world and is of particular concern with respect to groundwater protection.

FIGURE 2.5 Refinery complex with oily waste disposal pit in the foreground. The oil pit has since been remediated and no longer exists. (Photo courtesy of Komex.)

It moves often only a few meters a year, unseen and unsuspected below ground, until one day it "suddenly" appears meters or kilometers from the original source, in a water well or river, or discharging to a wetland. At that point, the race is on to control the damage and to find out where it came from. Once a problem has come to light, the next step is usually to determine the source and, in the "polluter pays" regime in which much of the industrialized world operates, assign blame. In practice, this can be very difficult. Tracing back part-per-million or even part-per-billion levels of specific pollutants through an aquifer that may be tens of meters below ground is a highly technical and, unfortunately in many cases, expensive undertaking. First, the direction and speed of groundwater flow need to be established. Hydrogeologists use many techniques, including drilling and sampling of exploratory wells, to develop a picture of the extent and concentrations of the offending chemicals. But the subsurface is not homogeneous. It is composed of a bewildering assortment of layers, fractures, and rock types. Each of these different features can deflect or even absorb contaminants. Usually, the best that earth scientists and environmental engineers can do is to come up with an incomplete view of the likely extent of contamination and some idea of its severity.

Assuming that we have determined roughly the what, where, when, and who of the problem, the next step is the hard part — fixing the problem. Unlike rivers or lakes, which have comparatively high rates of circulation and renewal, water moves slowly through the small fissures and pores of aquifers. Once a contaminant is introduced into the aquifer, it may take decades or even centuries to flush out completely. All this means that once it becomes contaminated, groundwater is generally very difficult and expensive to clean up.

In the simplest terms, this basic introduction to the problems of groundwater contamination provides the basis for application of economic techniques to help in better decision making. With so much at stake, in terms of both money and resources, it is important that the limited funds available for groundwater protection and restoration are allocated efficiently.

This is far from trivial. As described earlier, hydrogeology and groundwater remediation are relatively new disciplines and are inherently fraught with uncertainty. The science of environmental economics, particularly the valuation of the benefits of remediation, is also new and developing. Despite this combination of an inexact science with a dismal one, economics provides another tool to help make better decisions about how best to remediate problems of groundwater contamination.

2.2 THE REGULATORY PERSPECTIVE

When faced with a decision regarding remediation of a contaminated site or aquifer, a problem holder (private firm, organization, or individual) will almost always conduct its own financial analysis of the project, to determine whether to proceed. The anticipated costs of remediation will be compared to the benefits the firm expects to accrue, such as increased land value or reduction in corporate liability. This analysis is not strictly an economic one, because it considers only the costs and benefits of the problem holder, not of society as a whole. A whole range of other groups may have a stake in the remediation of the site, including neighbors, environmental groups, and owners or custodians of resources that may be impacted or are being impacted by the contamination. It is the role of the regulatory bodies to represent the interests of society as a whole, when considering contaminated sites and their remediation. As such, many jurisdictions have recently enacted legislation or guidance that calls for the full costs and benefits of remediation to be assessed as part of the decision-making process. In many places, remedial decision-making guidelines and legislation focus on protection or remediation of groundwater and sensitive and economically valuable aquifers. In either case, economic cost–benefit analysis (CBA) can be used to determine which of a number of remedial options will produce the highest net benefit for society as a whole.

2.2.1 U.S. REGULATIONS

The U.S. environmental legislative framework is regulated by the U.S. Environmental Protection Agency (EPA). More than a dozen major statutes and laws form the legal basis for the programs of the U.S. EPA. The EPA regulates at both a federal and a state level through a variety of state or tribal agencies (Cal EPA, Missouri Department of Natural Resources [DNR]). The National Environmental Policy Act 1969 established a broad national framework for environmental protection; it provided policy, goals, and mechanism for carrying out policy. Subsequent statutes and laws include the Clean Air Act (CAA 1970); the Safe Drinking Water Act (SDWA 1974); the Resource Conservation and Recovery Act (RCRA 1976); the Toxic Substances Control Act (1976); the Clean Water Act (CWA 1977); the Comprehensive Environmental Response, Compensation, and Liability Act (CERCLA, 1980); and the Superfund Amendments and Reauthorization Act (SARA 1986). The majority of contaminated sites are regulated at a federal or state level by CERCLA (and the 1986 amendment SARA) and RCRA.

2.2.1.1 Superfund

CERCLA, or Superfund, created a tax on chemical and petroleum industries and provided federal authority to respond directly to releases or threatened releases of hazardous substances. Over a five year period, $1.6 billion was collected into a trust fund for remediating abandoned or uncontrolled hazardous waste sites. CERCLA also established prohibitions and requirements pertaining to hazardous sites and provided for liability of responsible parties. CERCLA provided two response actions: a short-term rapid response and a long-term remedial response to releases that are not immediately life threatening. Amendments to Superfund in 1986 (SARA) first made the provision that remedial actions should be cost effective, but this has not been defined in strict economic terms. Clean-up priority criteria include:

- Affected population
- Specific health risks
- Potential for human contact
- Ecosystem impacts
- Damage affecting the food chain

Implicit in these criteria are an economic dimension, although this is not explicitly defined. Normally, however, Superfund calls for groundwater to be cleaned up to drinking-water standards (MCLs, or maximum contaminant levels). Furthermore,

Superfund provisions also allow state and federal governments to sue polluters for damages to natural resources, and stress the importance of permanent remedies and innovative treatment technologies.

2.2.1.2 The Resource Conservation and Recovery Act (RCRA)

The RCRA act was a significant modification of the Solid Waste Disposal Act of 1967 and provided the EPA cradle-to-grave authority over hazardous waste, including generation, transportation, treatment, storage, and disposal. RCRA also provided a framework for the management of nonhazardous waste. Subsequent amendments in 1986 enabled the EPA to address environmental problems relating to underground storage tanks storing hazardous substances. RCRA, unlike CERCLA, specifically focuses on active and future facilities and does not address abandoned or historical sites.

2.2.1.3 Enforcement

Under CERCLA and RCRA, the federal EPA and local state agencies are able to implement MCLs in soil or groundwater through enforcement measures, which can include fines of up to $25,000 per day of regulatory noncompliance. The U.S. environmental legislation and regulatory enforcement are based on health criteria. Remedial goals are based on reducing the concentration of a contaminant of concern (COC) below the MCL (health) for that COC, eliminating the pathway of COC to receptor (risk) and a cost–benefit analysis of the appropriate remedial technology to achieve the remedial goal (economics). At no stage in the process of legislation to remedial action is a value attached to an individual's health. The economics of contaminated site remediation operates at a policy-formation level and in the formulation of a range of incentives either to prevent pollution or to remediate pollution once a release has occurred. In the past decade, the EPA has started to use a broader range of environmental management tools than previously. Traditional regulatory systems provide an incentive to comply by avoiding enforcement actions, but the release of pollution itself incurs no economic cost. No incentive is provided to do more than is required within the regulations, be that an emission limit or specific technology. An economic value can be applied to pollution sources through market incentives, producing public health, environmental, and economic benefits. Incentive systems generally fall into the following categories:

- Pollution charges, fees, and taxes
- Deposit–refund systems
- Marketable permit systems
- Subsidies for reducing pollution or improving the environment
- Liability for harm caused by pollution
- Information disclosure
- Voluntary pollution reduction programs

One recently enacted example is the Small Business Liability Relief and Brownfields Revitalization Act 2002. The legislation is designed to provide relief from CERCLA liability for small business. The legislation has also been designed to provide financial assistance for brownfield revitalization and for promoting the clean-up and reuse of brownfields.

2.2.1.4 Groundwater Protection and Remediation

In the U.S., protection of groundwater resources is one of the major drivers of remedial activity. Groundwater quality protection law in the U.S., according to the NRC, is highly fragmented.[2] There is currently no single unifying federal ground-water quality protection law; at the state level, groundwater legislation is far from uniform. The federal Clean Water Act (CWA) does not cover groundwater quality the way it does surface water. Federal groundwater quality protection programs are scattered throughout a variety of federal laws.[2] Valuation of groundwater, as part of an effort at economic analysis for decision making, is included in cost–benefit analyses of Superfund clean-up alternative analysis, in evaluations of damage to natural resources, and in establishing new drinking water standards (MCLs) under the 1996 Safe Drinking Water Act. Relevant legislation addressing groundwater quality restoration in the USA includes:

- The federal Superfund program of the Comprehensive Environmental Response, Compensation, and Liability Act (CERCLA), which requires remediation of groundwater contaminated by waste disposal. Remediation may be required to MCL standards, but other clean-up targets may be approved by the EPA on a case-by-case basis. Amendments to Superfund in 1986 (SARA) first made the provision that remedial actions should be cost effective, but this has not been defined in strict economic terms. Clean-up priority criteria include (1) affected population, (2) specific health risks, (3) the potential for human contact, (4) ecosystem impacts, and (5) damage affecting the food chain. Implicit in these criteria are an economic dimension, although this is not explicitly defined. Normally, however, Superfund calls for groundwater to be cleaned up to drinking water standards (MCLs). This controversial requirement has brought criticism that real risks, and their full economic implications, are not accounted for in this approach. However, the EPA may relax groundwater standards on a site-by-site basis, reducing health risk standards from 10^{-6} (one additional cancer out of one million persons exposed over 70 years) to 10^{-4}. This ability to trade off remedial cost for increased risk is an implicit recognition that in some cases, the cost of remediation may not be warranted by the benefits (however expressed). Recently, the USEPA has started to grant technical impractablity (TI) waivers for groundwater remediation, in cases where existing technologies are clearly unable to meet remediation targets. Nevertheless, Superfund provisions also allow state and federal governments to sue polluters for damages to natural resources.

- The Resource Conservation and Recovery Act (RCRA) regulates hazardous waste transport and disposal, including USTs (underground storage tanks).
- Private clean-up liability exists under many state Superfund laws, under which polluters may be liable for the costs of groundwater remediation. State laws vary considerably with respect to the degree of financial responsibility borne by polluters.[2]
- The Clean Water Act (CWA) 1996 regulates point-source discharges into United States waters, including streams and wetlands, but not to groundwater. The CWA does not directly regulate non–point source pollution.
- The Safe Drinking Water Act (SDWA) 1996 establishes MCLs for public water supply and allows for remediation of contamination that results in exceedances of MCLs.
- Regulatory impact assessments (RIAs) were initiated by Presidential Executive Order 12291, requiring the EPA and other agencies to balance expected environmental protection with regulatory compliance cost, through preparation of cost–benefit analyses for proposed regulations imposing total costs of more than $100 million annually. The overall effect of RIA rules would be to "discourage rules for which a positive benefit–cost analysis cannot be generated or is marginal".[2] (Interestingly, the Canadian Council of Ministers of the Environment [CCME] now requires that all new environmental regulations and criteria in Canada be assessed on the basis of an economic RIA. This includes assessments of the effect on GDP and the "distributional" effects on various stakeholders.) The same requirement for an RIA at the policy level also exists in the U.K.

2.2.2   U.K. REGULATIONS AND GUIDANCE

2.2.2.1   Groundwater and CBA

In the U.K., both groundwater remediation and site remediation more broadly must be completed in a way that is economically defensible. The U.K. Environment Agency (EA) is required by law to consider the issues that are relevant to remedial works undertaken under Section 161a of the Water Resources Act 1991, as amended by the Environment Act 1995, and the general provision of the Environment Act 1995. In particular, the Agency must take into account the likely costs and benefits of remedial action when making its determination on a particular site or situation. To this end, the EA in the U.K. has developed guidance for remediation projects, including:

- *Review of Technical Guidance on Environmental Appraisal* is a report by the then department of environment transport and regions (DETR).[3] It is a discussion document for statutory guidance on contaminated land enshrines the role of risk assessment in developing appropriate solutions for contaminated land and broadly defines what is acceptable and not acceptable in terms of risk. The need to consider costs and benefits is included under Part III, "Reasonableness of Remediation."

- *Costs and Benefits Associated with Remediation of Contaminated Groundwater: A Framework for Assessment* is an environment agency draft supporting guidance on the application of cost–benefit analysis for groundwater remediation decision making.[4]
- *Cost–Benefit Analysis for Remediation of Contaminated Land* is an environment agency draft supporting model procedure for a study into land contamination and covers groundwater to some degree.[5]
- *Integrated Methodology for Derivation of Remedial Targets for Soil and Groundwater to Protect Water Resources* is an environment agency draft.[6]
- *Handbook of Model Procedures for the Management of Contaminated Land*, CLR-11, series, includes procedures for risk assessment and selection of remedial measures.[7]
- *Economic Valuation of Waste and Water Investments* is a review of technical guidance on environmental appraisal.[8]
- *A Methodology for Deriving Groundwater Clean-up Standards*[9] and *Methodology to Determine the Degree of Soils Clean-up Required to Protect Water Resources.*[10]

2.2.2.2   Brownfield Redevelopment

The U.K. has seen a significant movement to encourage redevelopment of brownfield land, much of which is a source of groundwater contamination (Figure 2.6). This has been prompted in part by the increasing pressures on green sites and countryside, as the demand for housing and commercial space grows. The U.K. currently has two tax relief schemes, five public-sector funding schemes, and a number of possible grants that can aid the development of brownfield sites.

The most recent fiscal regime aid to brownfield development is a scheme for reducing the tax burden of companies engaged in contaminated land remediation. This was introduced by Schedule 22 of The Finance Act 2001. For the purposes of corporation tax, land remediation expenditure shall be allowed as a deduction in computing the profits of the trade for the accounting period in which that expenditure is incurred. The wording of the act precludes the original polluter from this tax relief — the definition of the original polluter for tax purposes being the entity "in control of the land" (i.e., those who managed the pollution). This leads to the example that in the instance of purchasing land from a "polluting" company, there may be a benefit to both parties if the site is bought "dirty." In this way, the purchasing company would be able to benefit from tax relief for remediating the site, whereas the polluting company would not.

What the scheme means is that companies directly developing or investing in contaminated land can claim the additional costs of remediation and a 50% rebate. If the company does not have the profits to offset these allowances, they can claim a tax credit equivalent to 24% of the actual cost from the treasury. There is no

FIGURE 2.6 Reclamation of contaminated land — United Kingdom.

restriction on the type of development, and the contamination does not need to be causing harm — the risk that it may do in the future is sufficient.

The second fiscal regime aid is the Landfill Tax exemption scheme. The Finance Act 1996, which introduced landfill tax from October 1996, provides that waste materials disposed of by way of landfill constitute a taxable disposal. Section 43A of the Act deals with contaminated land and provides that there is no taxable disposal if a contaminated land exemption certificate relating to the materials removed has been granted. This has been one of the more effective methods of encouraging brownfield redevelopment. The Landfill Tax is a fiscal instrument, administered by HM Customs and Excise, and the exemption for contaminated land material means that brownfield developers avoid paying the tax for historically contaminated material. Developers and consultants often assess Landfill Tax liability as part of a prepurchase calculation of brownfield remediation and seek to pass this on to landowners through reduction in the purchase price.

Public-sector funding schemes are managed by English Partnerships (EP), the Government's national regeneration agency, which works with the Regional Development Agencies (RDAs) to bring private-sector investment to priority areas and to support the key regeneration projects identified in the RDA regional economic strategies. RDAs and EP are able to support a wide range of land and property regeneration projects via European Commission–approved schemes designed to regenerate derelict, disused, and vacant land around the country. The new schemes are funded by the Regional Development Agency's Land and Property budget, worth £1.55billion in 2002–2003 and £1.7billion in 2003–2004.

REFERENCES

1. API (American Petroleum Institute), *Underground Spill Cleanup Manual*, Am. Pet. Inst., Washington, DC, Pub. No. 1628, 1980.
2. National Research Council (NCR), *Valuing Groundwater*, NRC, Washington, DC, 1997.
3. DETR, *Review of Technical Guidance on Environmental Appraisal*, report by eftec, DETR, London, 1998.
4. Komex and eftec, *Costs and Benefits Associated with Remediation of Contaminated Groundwater: A Framework for Assessment*, U.K. Environment Agency Technical Report P279, Bristol, 2001.
5. Environment Agency, *Cost–Benefit Analysis for Remediation of Land Contamination*, Report No. TR P316, Bristol, 1999.
6. Environment Agency, *Integrated Methodology for the Derivation of Remedial Targets for Soil and Groundwater to Protect Water Resources*, R&D Publication 20, prepared by Aspinwall and Co., Bristol, 1999.
7. Environment Agency and DETR, *CLR-11 Handbook of Model Procedures for the Management of Contaminated Land* (draft), Environment Agency, Bristol, 1999.
8. eftec, *Economic Valuation of Waste and Water Investments*, Environment Agency, London, 1998.
9. Environment Agency, *A Methodology to Derive Groundwater Clean-up Standards*, R&D Technical Report P12, by WRc plc., Bristol, 1996.
10. Environment Agency, *Methodology to Determine the Degree of Soils Clean-up Required to Protect Water Resources*, Report P13, Bristol, 1996.

3 Groundwater Contamination, Risk, and Remediation

3.1 INTRODUCTION

All human activities may result in contamination of the terrestrial environment and the groundwater that lies beneath. Industrialization of many parts of the world over the past century has resulted in a large number of contaminated sites covering many millions of hectares.[1] Over the past two decades, particularly in North America and Europe, there has been a realization that contaminants introduced into the subsurface can cause severe environmental impacts, degrade valuable natural resources such as surface and groundwater, and adversely affect human health.[2]

This chapter seeks to provide a brief introduction to the key issues relating to contamination of land and groundwater, and develop a basic understanding of some of the key contaminant types and characteristics, investigation techniques, and practical remediation methods. This will set the context for a more detailed examination, in the remaining parts of the book, of remedial decision making and the use of economic analysis as part of that process.

The purpose of this chapter is to introduce the fundamentals of contaminant behavior in the subsurface (particularly organic contaminants such as hydrocarbon fuels and organic solvents), practical site investigation techniques, and effective remedial approaches. These are the tools all professionals, whether working with industry, with regulatory bodies, or as consultants, need in order to answer the fundamental questions of contaminated land and groundwater: What are the problems? What are the risks posed by those problems? Are those risks acceptable? If not, what should we do to eliminate those risks?

3.2 THE ISSUES

3.2.1 CONTEXT

Contamination of the subsurface environment may occur through a variety of mechanisms and may involve myriad different compounds with widely varying physical, chemical, and toxicological properties. Contamination of soil and groundwater may:

- Impact human health
- Degrade the environment, including ecological systems
- Damage natural resources
- Contaminate surface and groundwater resources

- Render land unfit for reuse
- Cause public concern
- Do financial harm to individuals, corporations, and governments
- Erode quality of life

Should any of these effects be deemed unacceptable, either to society as a whole, to landowners, or to corporations, some form of action may be required. The benefits of remediation can include:

- Protection of human health
- Prevention of further damage to natural resources and the environment
- Safeguarding of valuable water resources, such as aquifers and rivers
- Reuse of derelict land, preventing unnecessary development of green-field sites
- Increased land values
- Assuagement of public concern

Remediation, however, can be expensive. Deciding which sites need attention (or whether a particular site needs remediation) requires careful evaluation of many, sometimes conflicting, issues. First, is the site contaminated at all? If contamination is suspected, based on site history or knowledge of past activities, to what degree has the subsurface been impacted, and with what? Next, the dangers associated with the contamination need to be assessed. Are they unacceptable, based on current guidelines, regulations, and societal and corporate values? What are the economic realities of the situation? Can those involved and responsible pay for the required level of remediation? What level of remediation is justified? Who is responsible? And finally, if all of the answers lead to the need for remediation, and if the funds are available, what should be done? Which of many available remedial approaches and clean-up techniques should be applied, how, and by whom? When is the job complete?

3.2.2 Sources of Contamination

Contamination of land can be produced by almost any human activity. As the world has developed inexorably, planet-wide production of potentially harmful products, chemicals, and wastes has increased dramatically. Historically, groundwater contamination has resulted from a wide variety of industries and activities, including:

- Manufacturing
- Agriculture — farming, processing, production (Figure 3.1)
- Power generation
- Petrochemical refining, distribution (Figure 3.2), storage, and sale
- Petroleum exploration and development
- Mining and smelting (Figure 3.3)
- Weapons and explosives
- Coal gasification

FIGURE 3.1 Agriculture feeds the world but also affects groundwater in many parts of the world. Widespread use of fertilizers, pesticides, herbicides, and other chemicals may lead to deteriorating groundwater quality. (Courtesy of B. Tompkins.)

FIGURE 3.2 Petrol service stations are a common sight at street corners and along roadways the world over. Fuel stored in underground tanks may leak into surrounding soils and eventually impact groundwater. (Courtesy of Komex.)

FIGURE 3.3 Open pit mining operations can generate metal-rich acidic leachates that may affect surface and groundwater. (Courtesy of Komex.)

- Wood treating and preserving
- Automotive maintenance, parts, and service
- Nuclear energy, systems, and weapons

The processes that can lead to introduction of unwanted compounds into the subsurface include:

- Leaks and releases from processing and manufacturing facilities
- Leaks and spills from pipelines and storage tanks (Figure 3.4)
- Spills from tanker trucks and railcars
- Leaks and seepage from landfills
- Flow from pits, ponds, basins, and other structures built to hold wastes
- Uncontrolled dumping and fly-tipping
- Seepage from historical waste dumps, tailings ponds and piles, tips, and stockpiles
- Direct application (or overapplication) for other purposes (such as crop spraying and fertilization of farmland)

Despite a significant body of experience and research in the field of contaminated land (particularly in the U.S.), much of the world continues to degrade and pollute its natural environment at an alarming rate. The lessons learned in the U.S. and Europe are being largely ignored in many developing nations in the push for development and prosperity. Unfortunately, the increasing pace of global development is producing an unwanted legacy of polluted soil and groundwater.

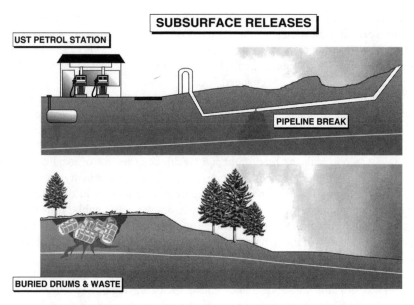

FIGURE 3.4 Schematics of subsurface releases to groundwater.

Perhaps the most important lesson of the past 30 years is that the best way to deal with contaminated soil and groundwater is to prevent contamination in the first place. The costs of remediation are commonly two or more orders of magnitude greater than the costs of prevention, and in many situations, full restoration of polluted soils and groundwater is not technically achievable.

3.2.3 DIFFERING STAKEHOLDER PERSPECTIVES: INDUSTRY, GOVERNMENT, AND SOCIETY

Many groups have a stake in the contaminated land issue. Each brings its own views, priorities, and agenda to the debate. Effective solutions require careful consideration of all views.

The land owner has immediate and direct involvement. However, the land owner may not necessarily be responsible for contamination. Tenants may have engaged in activities causing ground pollution, impacting the landlord's property. Contaminant may have migrated onto the site from an adjacent property. In these cases, establishing the environmental conditions at the site at the time of purchase or at the start of a lease may protect the property owner from liability.

In many parts of the world, including the U.K., land purchase is subject to the maxim *caveat emptor* (buyer beware). A purchaser is responsible for the contamination acquired, unless specifically exempted in the transaction documents. The onus is on the purchaser to satisfy himself that the liabilities are known and accounted for. Today in the U.K., Europe, and North America, and increasingly in the rest of the world, property transaction is preceded by a full environmental review of the liabilities associated with the property. The cost of remediation is often set against the purchase price, based on negotiation between the purchaser and seller.

Understanding the nature, scope, and seriousness of contamination at a site requires specialist knowledge and training. Most firms and organizations do not hold such expertise in-house, so *specialist environmental consultants* are called in to assist. Credentials, experience, and references should be checked carefully before consultants or contractors are engaged.

The *regulatory authorities* play an important role in the process. In the U.S., the United States Environmental Protection Agency (USEPA) works with state environmental agencies to regulate issues of contaminated sites and groundwater. In the U.K., the Environment Agency (EA) holds national responsibility for contaminated land issues, along with the Environmental Health departments of local councils.

The public (which may include interested individuals, community organizations, and action groups) has an increasingly important role to play in decision making regarding contaminated land. Many jurisdictions now require public consultation as part of a major contaminated land remediation program. Including a broad spectrum of interests ensures that the remedial objectives and approach selected meet society's broad objectives.

A cooperative approach to contaminated land issues is strongly recommended. The widest possible range of stakeholders, including regulatory bodies and the public, should be involved early on and kept engaged throughout the process. One of the problems with multistakeholder involvement in decision making has typically been that different groups' concerns are expressed and valued differently, using different units of measure. The problem holder's point of view is often crystallized as hard costs for remediation, based on well established remedial methods and cost databases. However, a local action group interested in preserving a nearby site of ecological value being impacted by the contamination would present its concerns in ecological and personal-preference terms. When a final decision comes to be made, the concerns of those stakeholders who have not expressed their position in terms of money are often relegated to the position of a footnote in a report, or a vague, emotion-laden statement about protecting the aesthetics and ecology of the area. How do you put a value on aesthetics? How does one describe the benefits of a thriving wetland in terms that a chief financial officer will understand? This book offers a way of doing just that, *allowing all stakeholders' views to be measured and compared in a common unit of value that everyone understands: money.*

3.3 CONTAMINATION OF THE SUBSURFACE ENVIRONMENT — AN OVERVIEW

3.3.1 SOILS, ROCK, GEOLOGIC STRUCTURE

The subsurface is most commonly composed of natural geologic materials — soils, unconsolidated sediments, and rock deposited and reworked over time. An in-depth understanding of subsurface contamination requires a sound background in geology, geomorphology, and sedimentary processes. Basic introductions to these subjects abound.[3]

In developed urban environments, the shallow subsurface may include *man-made ground* (fill) and can contain many buried structures (foundations, sumps, mine workings). These materials may also become contaminated and will need to be considered. Fill materials are common in cities and towns throughout Europe, for instance, and commonly include construction debris (brick, rubble), dredged silts, domestic rubbish, clinker and ash from coal processing, foundry slag, and mine waste rock. These materials are usually highly heterogeneous, both vertically and areally, and difficult to characterize. They rarely exceed 10 m in depth.

Natural geologic materials can be characterized according to their origin. Of major interest in the study of contaminated soil and groundwater is the *near-surface environment*, consisting of deposits that lie within 100 m or so of the surface. Spills and leaks originating from surface processes will impact near-surface materials predominantly. Deeper migration is possible and, although less common, does occur.

Soils and unconsolidated deposits cover much of North America and Europe. These materials are typically readily disaggregated and have considerable porosity (proportion of total volume that is void space) and permeability (ability to transmit fluid flow). Typical surficial deposits include materials of glacial origin (till), lacustrine sediments (typically fine grained and organic-rich), fluvial deposits (often composed of permeable sands and gravels), and aeolian wind-blown deposits. Typically, high-energy depositional environments (such as rivers) will produce more coarse-grained and permeable deposits. Finer-grained sediments (more clay- and silt-rich), are typically less permeable and contain higher proportions of organic material. The geologic origin and chemical makeup of the individual grains will have an effect on groundwater chemistry and contaminant behavior. Unconsolidated deposits typically are characterized by structure across all scales, including preferential grain orientation (imbrication), layering, faulting, and reworking features. The properties of geologic materials are important factors for the subsurface behavior of contaminants.

Consolidated materials (bedrock) are close to the surface in many parts of the world. These include consolidated sedimentary rocks such as sandstones and carbonates, metamorphosed sediments, volcanic deposits, and intrusive igneous rocks. Many important major aquifers in the United Kingdom, for instance, such as the chalk and the Triassic sandstones, are bedrock formations. Important characteristics for subsurface contaminant behavior include the common presence of fracturing (especially near surface), faulting, and matrix porosity and permeability (primary and secondary), all of which can provide pathways for contaminant movement (Figure 3.5 and Figure 3.6). A detailed understanding of any subsurface contamination problem should begin with a thorough review of the regional and local geology and an analysis of the parameters that may affect contaminant behavior and mobility.

3.3.2 THE HYDROLOGIC CYCLE

Much of the concern over contaminated sites arises from the potential for pollution of groundwater and surface water. Many common contaminants, including many organic chemicals, metals, and salts, are water soluble to varying degrees. This

FIGURE 3.5 Silurian dolomite in the United States. Fractured sedimentary rocks like this can make excellent aquifers, with fractures providing enhanced permeability.

allows potentially harmful compounds to be leached by rainfall and flushed through unsaturated ground, and to flow with groundwater. The mobility imparted by the hydrologic cycle provides an important mechanism for contaminants to reach *receptors* of concern. An understanding of the principles of hydrology and the hydrologic cycle is vital in the development of investigation programs, the evaluation of risk, and the planning and execution of remediation and mitigation schemes.

Figure 3.7 provides a basic overview of the hydrologic cycle. *Precipitation* in the form of rain or snowfall reaches the surface. Snow and ice accumulate as storage in cold months and are released as the temperature warms. Part of the water infiltrates into the ground and is taken up as soil moisture and *recharge* to groundwater. Some of the water runs off to wetlands, streams, rivers, and lakes. Some is evaporated and

FIGURE 3.6 Silurian dolomite. A knowledge of geology and geologic processes is an important part of understanding the behavior of contaminants in groundwater.

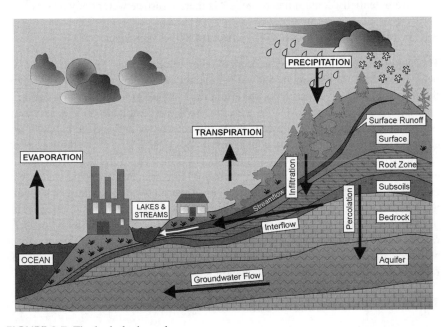

FIGURE 3.7 The hydrologic cycle.

transpired by plants. Groundwater flows and discharges to surface water bodies as *baseflow*. A balance between inputs, outputs, and storage is maintained, termed a *water balance*. Each part of the cycle is governed by a complex set of interactions and is the subject of a considerable science and literature. Reference should be made to texts such as Chow et al.[4] for a thorough discussion of hydrology and the hydrologic cycle. The key elements of the cycle with respect to contaminated sites are:

- Recharge — Infiltrating rain water and snow melt will leach water-soluble contaminants from soil, carrying them away as a dissolved phase. This may allow shallow soil contamination to have an effect on relatively deep groundwater or to affect nearby surface waters.
- Groundwater — Groundwater is a vital resource in many parts of the world, and potable aquifers require protection from contamination. Shallow groundwater is often of very poor quality. However, because groundwater flows (at a direction and speed which may vary temporally), contaminants in low-quality shallow zones may migrate and eventually reach a more sensitive and important water body.
- Surface water — Lakes, rivers, wetlands, estuaries, and marine coastlines are some of the most important and sensitive parts of our environment. Uses including potable water supply, recreation, and support of natural habitat may be degraded or destroyed by even low levels of contamination. As such, surface water bodies are key elements of study in any contaminated land problem. One of the first questions to ask when considering a potentially contaminated site is: "Is there a surface water body nearby?"

3.3.3 HYDROGEOLOGY AND GROUNDWATER CONTAMINATION

The study of hydrogeology (sometimes called *groundwater hydrology*) and the behavior of contaminants in groundwater, are complex disciplines in their own right. Several comprehensive textbooks are available covering groundwater and contaminant hydrogeology in detail.[2,5–7] In addition, books are available on a wide variety of specialized groundwater subjects, including aquifer testing[8] and the behavior and characterization of dense nonaqueous phase liquids (DNAPLs).[9,10] In addition, a vast peer-reviewed literature has been accumulated over the last several decades, covering every aspect of the behavior of a wide variety of contaminants in groundwater, from organic compounds to radioactive substances. Accordingly, the intent of this section is not to provide a comprehensive review of the theory of groundwater contamination, but rather to direct the reader not already familiar with these subjects to useful references. For completeness, this section provides a brief overview of some of the main concepts of groundwater contamination, which will assist those readers less familiar with the discipline to understand the main thrusts of the book.

3.3.3.1 Groundwater Hydrology

The formal study of groundwater hydrology dates back to the early part of the 19th century. In 1856, Henry Darcy published a treatise on the flow of water through porous media, based on experiments done while working on public water supply development in the south of France.[11] What has become known as the *Darcy equation* now forms the basis of our understanding of groundwater flow. Darcy related the flow (Q) through a porous medium of known cross-sectional area (A) to the applied head difference (Δh) over a given distance (Δx) and a constant that was the function of the properties of the porous medium (hydraulic conductivity, K) as in Equation 3.1:

$$Q/A = K \,(\Delta h/\Delta x) \tag{3.1}$$

Q/A is also known as the *Darcy velocity* (q), and hydraulic conductivity (K) is defined as in Equation 3.2:

$$K = \frac{k\rho g}{\mu} \tag{3.2}$$

where k is intrinsic permeability of the medium, μ is dynamic viscosity of the fluid (water), ρ is density of the fluid (water), and g is gravity. Hydraulic conductivity is usually expressed in units of length/time (m/s), and varies in geologic media from about 10^{-2} m/s, for a clean gravel, to 10^{-10} m/s for crystalline basement rocks. Freeze and Cherry[2] present a widely used table of common geologic materials and their typical hydraulic conductivities.

Darcy's equation is based on the fact that groundwater flows from high total head to low total head, where hydraulic head (h) for typically slow-moving groundwater is defined as the sum of the elevation head above a given datum (z) and the pressure head (P), and is expressed as in Equation 3.3:

$$h = z + (P/(\rho g)) \tag{3.3}$$

Hydraulic head provides the energy for groundwater flow. Figure 3.8 shows a schematic representation of groundwater flow from high to low total head.

In its basic form, the Darcy equation also represents geologic media as having a single hydraulic conductivity value K. In fact, the subsurface is infinitely varied and highly heterogeneous on all scales. In most instances, this heterogeneity defies

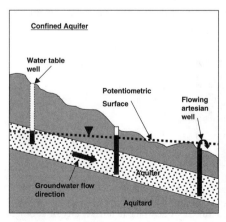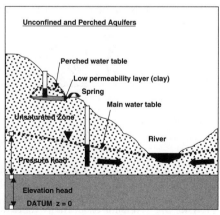

FIGURE 3.8 Schematics of groundwater flow.

practical description, so hydrogeologists will typically assign a single bulk equivalent value for K. However, K can be expressed in three dimensions (Figure 3.9), expressing the anisotropy (variation depending on direction) imparted by heterogeneity.

Groundwater occurs throughout the subsurface. Geologic strata containing usable groundwater (usually of a reasonable quality) that can flow to wells, and is thus suitable for practical exploitation, are termed *aquifers*. For the purposes of this book, aquifers are of major interest and can be described as having economic value in their own right. Finer-grained, lower-permeability strata that bound or isolate aquifer are called *confining beds*, or *aquitards*.[6] For more details on aquifer characterization, management, and behavior, see Freeze and Cherry and Schwartz and Zhang.[2,6]

3.3.3.2 Contaminant Hydrogeology

Contaminant hydrogeology encompasses the disciplines of hydrogeology, hydrology, and environmental geochemistry. It is the study of the way contaminants behave in groundwater. Groundwater flows in the subsurface at velocities that typically range from a few centimeters to several meters per year. The geologic media through which groundwater travels are usually very complex and heterogeneous, making detailed prediction of groundwater and contaminant behavior difficult. Estimating groundwater flow velocities, contaminant travel times, and similar parameters is subject to significant uncertainty, stemming in part from the considerable heterogeneity of subsurface materials. For instance, hydraulic conductivity (K), which describes the ability of groundwater to flow through a porous medium, can rarely be measured in the field to an accuracy greater than about half an order of magnitude. All groundwater calculations are subject to inherent uncertainty.

A wide variety of common products have the potential to contaminate groundwater. Common contaminant types include hydrocarbon liquids, organic solvents, heavy metals, inorganic compounds, fertilizers, pesticides, herbicides, and radioactive compounds.

Once in the subsurface, concentrated accumulations of contaminants may remain for long periods of time, bound to or trapped within the soil or rock, essentially

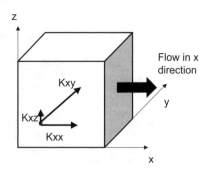

FIGURE 3.9 Hydraulic conductivity varies directionally due to the heterogeneity of geologic materials.

immobile. These are commonly termed *subsurface sources*. If any of these contaminants has an appreciable solubility in water, dissolved phase contamination will be produced from the subsurface sources, and it will then migrate away with groundwater flow.

Contaminants in groundwater are subject to various physio-chemical and biological processes that will effectively retard their movement or reduce their concentration over time. These include adsorption onto geologic materials, biodegradation, chemical breakdown, dilution, and dispersion.

Adequate site investigation is critical for providing the data with which to understand the groundwater flow regime, delineate contamination, and identify the types and concentrations of contaminants. Prediction of the rates and patterns of contaminant movements in groundwater is not always straightforward and is subject to considerable uncertainty, despite the advent of sophisticated computer modeling techniques. Sensitivity analysis is often used to explore the likely possible range of predictions.

Contaminants that dissolve in groundwater are then available to migrate with groundwater as it flows in the subsurface. Infiltrating precipitation and groundwater can dissolve water-soluble compounds, which are carried away by groundwater flow. Dissolved phase plumes are subject to hydrodynamic dispersion, which may cause spreading, and retardation due to adsorption or biological/abiotic transformations. These factors tend to slow the apparent migration of reactive compounds (such as organics), compared to that of a conservative nonreactive solute, such as chloride. Again, numerous specialized references on the subject of dissolved phase contaminant transport in groundwater are available. The reader is referred to Freeze and Cherry, Schwarz and Zhang, and Fetter for excellent introductions to the basics of mass transport.[2,6,7] Schmelling and Ross[13] provide an overview of dissolved phase contaminant transport in fractured rocks.

As mentioned earlier, accumulations in the subsurface of contaminants in concentrated form (perhaps in a landfill, as buried wastes, or as spills of liquid hydrocarbons, for instance) are often termed subsurface sources, because they act as continuous long-term sources of dissolved phase contamination to groundwater. Although subsurface sources represent a concentrated contaminant mass, associated dissolved phase plumes are dilute, mobile, and often much farther ranging. The implications for remediation are obvious: removal of the concentrated source is often the most effective remedial strategy and often the most cost effective.[14,15] Time can also be an important factor: the longer we wait to remove the source, the farther the dissolved phase plume migrates, and the greater the volume of aquifer that it contaminates (Figure 3.10).

Contaminants in the aqueous phase (dissolved in groundwater) are moved by the bulk flow of groundwater. This process is known as *advection*. The average pore velocity (v) can be expressed as a modification of the Darcy equation, by dividing the Darcy velocity (q) by the effective porosity (n_e), yielding Equation 3.4:

$$v = q/n_e \qquad\qquad (3.4)$$

FIGURE 3.10 Removal of subsurface source of coal tar by excavation, as part of a remediation project at a manufactured gas plant in the United Kingdom. (Courtesy of Komex.)

Effective porosity describes that portion of a medium's porosity which is effectively available for flow, and does not include dead-end pores and sealed pores.

As groundwater flows through a porous medium, it follows a tortuous path. This leads to contaminant spreading and is known as *mechanical dispersion*. Figure 3.11 shows a schematic of a typical dissolved phase plume, so called for the characteristic featherlike shape from source at the stem, spreading as it moves down-gradient. Solute transport can be described mathematically by a series of partial differential equations, known as the *advection–dispersion equation*.[12]

As contaminants move in groundwater, they are also subject to a variety of physical, chemical, and biological processes that act in combination to retard or slow down the apparent movement of the plume. One of the most important of these processes is sorption. *Sorption* is the term used to describe the reaction of contaminants with the surfaces of the soil or rock matrix, and is generally considered to include adsorption (interaction at the surface of soil particles) and absorption (uptake of the compound within the soil matrix). Both mechanisms are believed to be important for organic contaminants.[16] This effect is often described as a *sorption isotherm*, which exists for a specific organic compound on a specific soil. For a *linear isotherm*, the partitioning coefficient K_d expresses the ratio of the equilibrium between sorbed and dissolved phase concentrations, and equals the slope of the isotherm (in units of mL/g) (Equation 3.5):

$$K_d = C_s/C_w \tag{3.5}$$

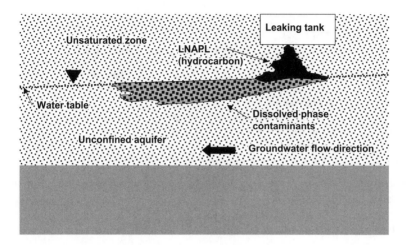

FIGURE 3.11 Light nonaqueous phase liquid (NAPL) hydrocarbon in the subsurface acting as a source of dissolved-phase contamination in groundwater.

where C_s is the concentration sorbed to the solid phase, and C_w is the concentration in the aqueous phase. This can then be related to groundwater flow velocity, and a *retardation factor*, R, is produced as in Equation 3.6:

$$R = 1 + (\rho_b/n_e) K_d \qquad (3.6)$$

where ρ_b is bulk density of the solid and n_e is effective porosity. R is used in the advection–dispersion equation as a retardation coefficient. When $R = 1$, no retardation occurs, as would be the case for a conservative solute (one that does not react with the matrix or degrade in any other way, such as chloride). Movement of solutes with higher values of R will be retarded (slowed down) compared to conservative solutes.

Sorption of organics to soil organic carbon is one of the most important components of sorption. In this case, adsorption becomes in part a function of the content of organic material within the soil matrix, or fraction of organic carbon (f_{oc}), measured as a percentage by weight. Generally, as f_{oc} increases, K_d increases. In cases where sufficient organic material is present, K_d is generally directly proportional to f_{oc}. In such cases, it is common to calculate a normalized sorption coefficient, K_{oc}, as in Equation 3.7:

$$K_{oc} = K_d/f_{oc} \qquad (3.7)$$

K_{oc} values are available in the literature for common organic compounds, and a selection of K_{oc}'s for common organic contaminants is provided in Table 3.1. The greater the K_{oc} is, the greater the compound's tendency to adsorb to organic material within the subsurface and the more highly retarded the plume. K_{oc} values in the

TABLE 3.1
Example of K_{oc} Values for Organic Compounds

Compound	K_{oc} (mL/g)	Specific Density
Dichloromethane	8.8	1.33
Trichloroethylene (TCE)	86	1.46
Benzene	83	0.87
Dichlorobenzene	1700	1.31

single digits are considered low, in the tens to hundreds moderate, and in the thousands highly adsorptive.

3.3.3.3   Nonaqueous Phase Liquids (NAPLs) in the Subsurface

Organic liquids such as gasoline (petrol), diesel fuel, and crude oil are among the most common groundwater contaminants.[17] In pure liquid form, organic liquids are referred to as *nonaqueous phase liquids* (NAPLs) and are defined as liquids that exist for some period of time as a separate, immiscible phase in water. Perhaps one of the most important aspects of NAPL contamination of the subsurface is the potential for creating long-term sources of dissolved-phase groundwater contamination. As groundwater or infiltrating precipitation flows past NAPL in the pore spaces, water-soluble species contained within the NAPL will partition into the aqueous phase and move off into the surrounding environment via groundwater flow. As long as NAPL remains in the ground, aqueous-phase contaminants continue to be released, sometimes over tens or hundreds of years.[18] This highlights one of the chief concerns over the presence of NAPLs in the subsurface and is a major impetus for remediation of NAPL spills.[15,19]

As such, NAPL behavior as groundwater contaminants are discussed in greater detail subsequently, as an example. Although it is beyond the scope of this book to go into every type of contaminant in detail, the following brief discussion is intended to highlight the level of complexity and study required to understand each type of contaminant that may affect groundwater.

NAPLs that have densities less than that of water, causing them to tend to float on a free water surface, are referred to as *LNAPLs* (light NAPLs). The most common types of LNAPLs are petroleum hydrocarbons and distillates, such as crude oil, gas condensates, gasoline, diesel and aviation fuels and lubricants. NAPLs with densities greater than water are referred to as *DNAPLs* (dense NAPLs) and include chlorinated solvents, PCBs, and heavy coal tars. NAPLs can also occur as mixtures of various dense and light compounds, with the resultant density (and thus behavior) dependent on the various proportions of each compound. These products are widely used in all modern societies and represent one of the major sources of groundwater and soil contamination in industrialized and developing countries.

Hydrocarbon liquids are made up of complex mixtures of individual components, the majority of which fall into three basic categories: paraffins, naphthenes, and aromatics.[20] Aromatic hydrocarbons are composed of at least one benzene ring consisting of six carbon atoms. These compounds are of special environmental significance, because they are partially water soluble, and some (such as benzene) are known carcinogens. A list of the more common hydrocarbons is provided in Table 3.2.

3.3.3.3.1 Physiochemical Properties

Immiscible liquids are described by several key parameters, including density, chemical composition, boiling point, solubility, and viscosity.

Separation of crude oil into its components is achieved through distillation. Different hydrocarbons have different *boiling points*, depending on the number of carbon atoms in their structure. Lighter compounds (smaller carbon number) have lower boiling points, and many (such as benzene) volatilize at room temperature. This property is of importance when considering mobility through the subsurface, partitioning into the vapor phase, and remediation, as will be discussed later. Table 3.2 provides information on the boiling points and other physical properties of selected hydrocarbons.

Density is the mass per unit volume of liquid, and in the case of LNAPLs is by definition less than 1 g/cm^3. The density of typical hydrocarbon liquids varies from 0.554 g/cm^3 for methane to 0.88 g/cm^3 for orthoxylene. Table 3.2 provides further

TABLE 3.2
Chemical and Physical Properties of Selected NAPLs

Compound	Density (g/cm³)	Solubility (mg/L)	Viscosity (cp)	Boiling Point (°C)
Normal Paraffins				
Propane	0.582	62.4	79.5 @ 7.9°C	−0.5
Hexane	0.659	9.5	3260 @ 20°C	69
Branched-Chain Paraffins				
2-Methylpentane	0.669	13.8	–	60
2-Methylhexane	0.679	2.54	–	90
Naphthenes				
Methylcyclopentane	0.749	42	–	72
Cyclohexane	0.778	55	1.02 @ 17°C	81
Aromatics				
Benzene	0.879	1780	0.652 @ 20°C	80
Toluene	0.866	515	0.590 @ 20°C	111
Orthoxylene	0.880	175	0.810 @ 20°C	142
Metaxylene	0.864	146	0.620 @ 20°C	138.9
Paraxylene	0.861	156	0.648 @ 20°C	138
DNAPLs				
Dichloromethane	1.33	20000	0.44 @ 20°C	41
111 Trichloroethane	1.35	1300	0.84 @ 20°C	113
Trichloroethylene (TCE)	1.46	1100	0.57 @ 20°C	87

examples. In the case of DNAPLs (dense NAPLs), densities are greater than 1 g/cm³, and liquids will sink through a water column.

Dynamic viscosity (μ) is the internal friction within a liquid that causes it to resist flow.[9] It is the result of molecular cohesion within the liquid. Less viscous NAPLs will tend to move through a porous medium more readily, since hydraulic conductivity K is inversely related to viscosity (see Equation 3.2). Viscosities of some common hydrocarbon components are listed in Table 3.2.

Aqueous *solubility* of NAPL contaminants is of special interest, because it describes the tendency of individual hydrocarbon components to enter the dissolved phase, where they are much more mobile and can move with groundwater. Solubilities for various hydrocarbons, listed in Table 3.2, vary widely. In the presence of NAPL mixtures, however, the effective solubility rule applies, where the effective solubility of a component i in an NAPL mixture, S_{ie}, is a function of the pure phase solubility S_i of the compound and the mole fraction of that compound represented in the mixture X_i, as in Equation 3.8:

$$S_{ie} = X_i S_i \qquad (3.8)$$

Thus, field solubilities tend to be less than 10% of NAPL pure phase solubilities, even if NAPL is present.[9]

All immiscible liquids have some tendency to partition into the vapor phase. This effect is governed by the *vapor pressure* of a compound. Volatile compounds, such as many chlorinated DNAPLs, have high vapor pressures and partition readily into the vapor phase. As with solubilities, the vapor phase concentration of a compound is a function of the composition of the liquid from which it is derived. The equilibrium vapor phase concentration (c) of a constituent can be estimated by Raoult's law (Equation 3.9):

$$C = X(P/RT) \qquad (3.9)$$

where X is the mole fraction of the compound within the mixture, P is the compound vapor pressure at temperature T (in degrees K), and R is the gas constant. This mechanism can be important in the subsurface because organic vapors may migrate long distances in the unsaturated zone and repartition into the aqueous phase, causing distant groundwater contamination.[21] In addition, the presence of organic vapors in the unsaturated zone may lead to exposure or risk in some circumstances.

3.3.3.3.2 NAPL Behavior in Porous Media

Much of the work done to date on the behavior of NAPL contaminants in the subsurface has concentrated on porous unconsolidated media, such as sands and gravels. This is partially the result of strong legislation enacted in the U.S. directed at leaking underground storage tanks (USTs). USTs are most commonly used in highly populated areas, where unconsolidated sediments at the surface allow easy excavation. The bulk of NAPL contamination problems that have been investigated in North America consist of isolated hydrocarbon spills occurring in shallow, unconsolidated sands, silts, and tills.[17,19]

The majority of NAPL releases occur at or just below the ground surface, the result of spills and leaks from buried lines, storage tanks, and holding ponds. Within the unsaturated zone, NAPL will migrate vertically downward, under the influence of gravity, saturating available pore space as it goes. As the wetting front of NAPL progresses downward, it leaves behind residual trapped liquids. In heterogeneous media, vertical migration and lateral spreading of NAPL within the unsaturated zone may be significantly affected.

If the NAPL spill volume is sufficient to overcome the residual soil retention capacity, migration will continue until the groundwater capillary fringe is reached. Here, LNAPLs (lighter-than-water NAPLs) will begin to accumulate. As volumes of floating LNAPL increase, lateral spreading may begin. LNAPL spreads along the groundwater surface, both as a result of the natural flow of groundwater and due to the creation of an LNAPL head. The groundwater surface may also be depressed by the weight of the floating LNAPL. DNAPLs, on the other hand, are denser than water and will continue to migrate downward through the water column, wherever there is sufficient permeability.

A conceptual schematic comparing LNAPL and DNAPL releases in a heterogeneous subsurface is provided in Figure 3.12. Note the vastly different behaviors of the light and dense liquids. The geometry of a spill will depend on the volume of NAPL, the area and time over which it was introduced into the subsurface, the physical properties of the NAPL and the subsurface, the position and temporal behavior of the groundwater surface, and the heterogeneity of the host medium. On a smaller scale, the variations in NAPL distribution within the subsurface will depend on the pore size distribution and the fluid pressures of each phase (air, water, NAPL).

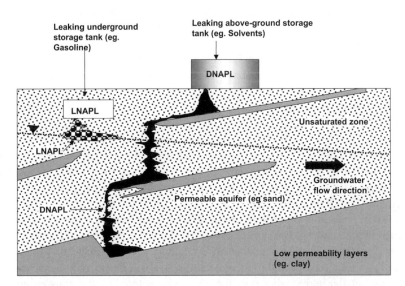

FIGURE 3.12 Light nonaqueous phase liquids (LNAPLs) and dense nonaqueous phase liquids (DNAPLs) behave differently in the subsurface. (Dissolved phase not shown for clarity.)

Once in the subsurface, NAPL movement is governed by the laws of multiphase flow. A detailed discussion of the complexities of multiphase flow in porous and fractured media is beyond the scope of this book, and the reader is referred to Mercer and Cohen and Pankow and Cherry for more information on the subject.[22,10] However, the basic properties governing multiphase behavior are discussed briefly here.

The *saturation* (*S*) of a fluid is defined as the proportion of available pore space that it occupies and may vary from 1.0 (fully saturated) to 0. In multiphase systems, such as NAPL and water, each phase competes for the available pore space, so all fluid saturations sum to 1.0. Field estimation of saturation is extremely difficult, even in porous media. Several methods have been used to determine *in situ* saturation, including pumping tests and geophysical logging; however, these remain qualitative in nature and largely undocumented.[22]

Wettability is the tendency for one fluid to spread over the solid phase surfaces of a multiphase system, coating the grains of a porous medium, while the other fluid occupies the remaining space.[23] In near-surface conditions, natural aquifers are strongly water wet.[22] In such systems, the wetting fluid will coat solid surfaces and occupy the smaller pore spaces and pore throats. The nonwetting fluid, usually the NAPL, tends to occupy the larger pores. Wettability is usually determined by the contact angle at the fluid–solid interface (ϕ).

Capillarity is the tendency for a porous media to draw in, or *imbibe*, the wetting fluid and repel the nonwetting fluid. In water–NAPL systems, capillary pressure is defined as the difference between the nonwetting fluid pressure and the wetting fluid pressure.[22] *Capillary pressure* is a function of the pore size, pore throat size distribution, and the interfacial tensions between the various phases. For an NAPL–water system, the interfacial tension between water and NAPL determines the capillary pressure that must be overcome for NAPL to enter a pore throat of radius *r*. Hence, finer-grained materials with small pore throats are more diffucult for NAPL (the nonwetting phase) to enter.

In reality, most geologic media are characterized by considerable heterogeneity, producing a wide range of pore sizes occurring across a spectrum of scales. At the field scale, for instance, sedimentary units are typically made up of interbedded layers of coarser- and finer-grained materials, such as sands, silts, and clays. Capillary pressure distributions and behaviors in each layer will be different, the result of different grain size distributions and pore structures. The effects of secondary alteration processes, such as weathering, fracturing, dissolution, and cementation, will further alter pore structure, sometimes cutting across original depositional boundaries. At the pore-scale, random variations in grain size, grain orientation, mineralogy, and imbrication lead to a complex distribution of pore throat and pore sizes. Hence, capillary pressure for a geologic medium cannot adequately be described by a single capillary pressure value. Rather, entry of NAPL into a water-saturated medium must be described by a capillary pressure distribution, or curve. At each pressure, only pore throats of a certain size will be entered. As the pressure increases, smaller and smaller pore throats can be entered by the nonwetting fluid. Hence, the capillary pressure characteristics of geologic media are among the most important factors governing mobility of LNAPL in the subsurface.

In a multiphase system, available pore space is occupied by more than one fluid. For an NAPL–water situation, water saturation reduces the available cross-sectional area for flow of NAPL, resulting in a lower effective permeability for the NAPL. This effect is known as *relative permeability*, which may vary from 1 (one phase flow) to 0 (residual saturation) for a given fluid. Hence, relative permeability is a function of saturation. Despite the importance of relative permeability to NAPL flow, and thus remediation, these types of data are rarely available for environmental investigations.[22]

LNAPL mobility may be enhanced by changing hydraulic gradients, or the properties of the fluid (by the use of surfactants, which decrease interfacial tension and reduce viscosity), or the properties of the medium (hydraulic fracturing to open up new flow pathways).[22]

Residual saturation (Sr) is the saturation at which NAPL becomes immobile within the porous medium. At this point its relative permeability is zero, and it is bound in place by capillary forces. Residual saturation is a function of the characteristics of the NAPL (density, viscosity, interfacial tension), wettability, pore distribution, and hydraulic gradients. At residual saturation, NAPL occurs as disconnected blobs and ganglions. Residual NAPL saturation usually varies from 0.1 to 0.5.[22,24] The immobility of LNAPL at residual saturation has important consequences in the characterization and remediation of hydrocarbon spills. Unable to flow as a separate phase, many NAPLs contain water-soluble components, which tend to partition into and move off with the surrounding groundwater. Although trapped NAPL can therefore result in significant and lasting dissolved-phase groundwater contamination, it is difficult to detect and remove. Bound by capillary forces, residual NAPL cannot flow readily to wells.

3.3.3.3.3 NAPL Flow

NAPL flow in porous media is governed by the same basic laws as groundwater flow but with the added complications of relative permeability–saturation effects. The movement of an NAPL from pore to pore within a porous medium is controlled by the capillary pressure characteristics of each successive pore, connected by pore throats. For NAPL migration to occur in the unsaturated zone, the pressure head of accumulated NAPL must be sufficient to overcome the entry pressure of the pores. Under hydrostatic conditions, the NAPL pressure head can be expressed as an equivalent connected height of NAPL. The critical height of NAPL required for pore entry (z_{nw}) can be expressed as Equation 3.10: [22]

$$z_{nw} = \frac{2\sigma \cos\phi}{r_i g \rho_{nw}}$$ (3.10)

where r_t is the pore throat radius, g is gravity, σ is the interfacial tension, ϕ is the wettability contact angle, and ρ_{nw} is the density of NAPL. Hence, the greater the accumulated vertically connected mass of NAPL, the greater is its potential downward mobility. As NAPL migrates deeper into the vadose zone from a continuous surface release, the driving pressure head actually increases, and the NAPL is able to enter pores with smaller and smaller pore throats. In the case of LNAPL, vertical migration is limited by the presence of the groundwater capillary fringe. Here, LNAPLs will begin to "float" and accumulate. DNAPLs, however, will continue to migrate vertically until a capillary barrier (a fine-grained or low permeability layer, with pores that are too small to enter) is reached.

3.3.3.3.4 NAPL Modeling

Modeling of LNAPL flow in shallow porous media is a relatively new science. Multiphase flow models were developed for petroleum industry applications in the 1970s and since then have been widely used by reservoir engineers to simulate hydrocarbon liquid flow within deep, high-temperature and high-pressure environments. However, the problems faced by workers in the environmental industry differ significantly in many ways. It has only been recently that models have been formulated to study near-surface LNAPL behavior.[25-28] In general, however, these models have been of a conceptual and research nature. Even in simple, relatively homogeneous media, data required to model LNAPL migration explicitly are most often lacking. This can be attributed to a variety of factors, including a lack of specialized field techniques, insufficient funds, and incomplete understanding of the dynamics of the subsurface processes involved.

3.3.3.3.5 Field Monitoring of NAPLs in Wells

The most widely practiced technique for assessment and monitoring of NAPLs is to install monitoring wells. A slotted pipe section and permeable annular material are placed across the interval of interest, sealing the rest of the section to surface. Groundwater and NAPL in the subsurface may then flow into the well, where samples can be collected and fluid elevations measured. Because mobile LNAPL usually occurs above the prevailing groundwater surface, monitoring wells should be carefully constructed. Perhaps the most commonly used indication of NAPL contamination is the presence of free NAPL within monitoring wells. Measurement of some thickness of NAPL on the surface of the standing water column within the well casing (in the case of LNAPL) or at the bottom of the well (for DNAPL) is a definite indication of the presence of NAPL within the formation itself. As such, it can be an extremely effective qualitative investigation technique. Using properly designed and constructed monitoring wells, investigators can delineate the approximate extent of the "free" hydrocarbon plume based on presence or absence of NAPL in monitoring wells.[29] However, if screens are placed below or above the prevailing groundwater surface, mobile NAPL cannot flow into the well, resulting in a "false negative": mobile NAPL exists at that point in the subsurface but cannot enter the well.

Many workers have attempted to relate the thickness of accumulated LNAPL in the well (the apparent thickness) to the actual thickness within the adjacent porous medium. Empirical evidence suggests that measured or apparent LNAPL thicknesses (ha) are typically two to ten times greater than the actual thickness within the adjacent formation (hf).[20,22] In the well, accumulated LNAPL depresses the groundwater surface. The capillary forces acting in the porous medium tend to act against groundwater surface depression. Thus, under equilibrium conditions, thicknesses would tend to be greater in the well than in the adjacent porous medium. Several theoretical and semiempirical relationships relating *ha* to *hf* have been proposed.[29-32] However, all have been found to be lacking, especially in their predictive capabilities.[33] One of the most widely used relationships was developed by de Pastrovich et al., which related the thicknesses in the well and the formation to the ratio of the densities of the two fluids (Equation 3.11):[34]

$$\frac{h_a}{h_f} = \frac{\rho_L}{\rho_w - \rho_L} \qquad (3.11)$$

where ρ_L if the LNAPL density and ρ_w is the density of water. Hence for a typical hydrocarbon fuel with a density of 0.8 g/cm^3, the apparent thickness would be four times the actual thickness in the adjacent formation. Although perhaps providing a useful rule of thumb, this and similar relationships are an oversimplification of the situation and may frequently provide erroneous estimates of actual LNAPL thickness, and thus volume. One effect that is not considered is the entrapment of LNAPL within the porous media as a result of groundwater surface fluctuations. As water tables rise, some LNAPL will follow, floating on the surface, but some will become trapped as the wetting fluid bypasses blobs and ganglia in larger pores. At this point, less LNAPL flows into the well. As groundwater surfaces drop, LNAPL drains from the unsaturated zone, accumulating and flowing into the well. An inverse relationship between groundwater surface fluctuation and LNAPL apparent thickness has been documented by Kemblowski and Chiang[30] and others. Some workers have now concluded that LNAPL apparent thickness measurements are not a consistently reliable indication of actual mobile LNAPL thicknesses or volumes.[14,29]

Another important limitation of NAPL thickness monitoring for description of an NAPL-contaminated site is the inability of the method to detect immobile NAPL. Residual saturations of LNAPL in porous media can be as high as 0.5. Hence, monitoring data may suggest no NAPL, although a significant volume of NAPL actually exists at the site. In these cases, high dissolved-phase concentrations (over about 1% of the aqueous solubility of the component) are usually taken as an indirect indication of the presence of NAPL.[9] Although this provides an inference of presence, it does not allow any quantification of the volume of NAPL present. NAPL behavior in fractured aquifers is an even more complex issue and is discussed as an example of applying economic methods to difficult remediation problems in Part IV.

3.4 RISK ASSESSMENT — GAUGING THE IMPACTS OF POLLUTION

3.4.1 BACKGROUND

Over the past two decades, the process of risk assessment has become one of the key elements of environmental decision making in European nations, Canada, the United States, and many other countries. Risk assessment explicitly recognizes that each situation involving contaminated soil or groundwater is different and may pose different risks to society. By assessing the risks posed by contamination, sites and problems can be prioritized and the most serious dealt with first. Risk assessment can equally be used to identify situations in which action is not warranted — contamination may exist, but it may not affect anyone to a degree that requires action. The process recognizes implicitly that not all receptors are as sensitive as others. Children, for instance, are among the most sensitive receptors (Figure 3.13). With low body mass and the tendency to play outdoors (and even consume soil [pica]), children can be exposed to contaminants more readily than adults, and the effects on them can be worse.

The results of risk assessment are a key input into an economic assessment. As will be described in more detail in Part II and Part III, quantification of the risks and impacts to receptors allows the benefits of remediation to be expressed in monetary terms. By preventing or mitigating identified risks, remedial action produces an economic benefit. Quantification of that benefit starts with understanding the impacts that would have occurred if the remediation had not taken place.

This brief overview of a complex subject is intended as an introduction for the nonpractitioner only. More detailed and comprehensive information on environmental risk assessment can be found in Linkov and Palma-Olivirea,[35] USEPA,[36] and Defra and EA.[37]

3.4.2 SOURCE, PATHWAY, AND RECEPTOR CONCEPT

Environmental risk assessment is based in the source–pathway–receptor concept. The potential for risk exists if there is a *source* of contaminants (a hazard), a sensitive

FIGURE 3.13 Risk exposure: children playing in the garden are the receptor. Ingestion of contaminated soil is the pathway.

receptor, and a *pathway* linking the two. A potential risk is said to exist only if all three (source, pathway, and receptor) exist. This is termed an *SPR* (or *risk*) *linkage*. Risk is defined as the probability that the receptor is adversely affected by the contaminant. End-point receptors may include controlled waters, humans, wildlife, ecosystems, buildings, and valuable resources. Figure 3.14 shows a schematic example of various ways in which SPR linkages can be created by groundwater contamination.

3.4.3 The Components of a Risk Assessment

Developing a risk assessment requires information from a number of sources, which form the basis of the analysis. A brief description of the necessary steps is provided in this section.

3.4.3.1 The Conceptual Model

The conceptual model provides the backbone to any risk assessment. It identifies potential sources, pathways, and receptors and determines the plausible risk linkages between them. Development of the conceptual model is iterative. A preliminary conceptual model should be developed prior to site investigation. This ensures that the investigation focuses on the primary pollutant linkages and most significant areas of uncertainty. Data from site investigation are then used to improve the conceptual model until a satisfactory understanding of the risks is achieved.

3.4.3.2 Site Investigation

Site investigation data provide the building blocks of the risk assessment. They are used to improve the conceptual model and eliminate any pollutant linkages that are unlikely to cause significant harm. Site investigation should also provide the necessary data for performing quantitative risk assessment, such as key retardation parameters for dissolved phase migration (fraction of organic carbon) and the key hydrogeologic parameters that govern solute transport (notably hydraulic conductivity).

3.4.3.3 Fate and Transport Modeling

The need for quantitative risk assessment is identified from the conceptual model. Fate and transport modeling is a necessary component of quantitative risk assessment. It determines the point of exposure contaminant concentration to which a receptor could become exposed. Various fate and transport modeling methods may be required, depending on the pathways involved. These include simple groundwater models, such as the U.K. Environment Agency's P20 spreadsheet[38] and more complex groundwater flow models, such as MODFLOW[39] and its derivatives and solute transport add-ons.

3.4.3.4 Toxicological Assessment

Toxicological assessment is also a necessary component of quantitative risk assessment. It is used to determine the acceptable contaminant dose or concentration to

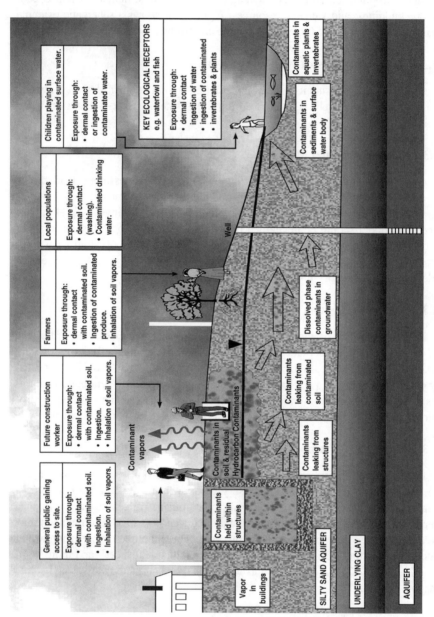

FIGURE 3.14 Conceptual schematic of sources, pathways, and receptors.

which a receptor can be exposed. These data can be used with the results of the fate and transport modeling to determine acceptable source concentrations at a site. In many cases, toxicological data have been used to determine generic point-of-exposure guideline concentrations, such as drinking water and environmental quality standards. Where these are not available or suitable for the pollutant linkages identified, site-specific toxicological assessment may be required. Toxicological impacts on humans and on ecological receptors may need to be considered.

3.4.3.5 Risk Quantification

Figure 3.15 shows a generic risk quantification process. There are many risk quantification methods available for assessing risks to humans, ecological receptors, and water bodies. In the U.K., the Department for Environment, Food, and Rural Affairs (Defra) and the Environment Agency (EA) have published the CLEA model.[37] The model consists of a series of algorithms for assessing the long-term risks to humans on contaminated sites. Pathways considered are ingestion and dermal contact of soil and dust, inhalation of vapors and dust, and ingestion of contaminated vegetables. Risc-Human is a computer model developed by the Van Hall Institute in Holland for assessing risks to human health. The model contains the CSOIL algorithms developed by the Dutch National Institute of Public Health and the Environment (RIVM). These were used for producing the Dutch soil and groundwater guideline values. Risc-Human allows assessment of risk from various exposure pathways, including ingestion; inhalation of and dermal contact with soil and water; inhalation of vapors; and consumption of meat, milk, vegetables, and fish that have been exposed to contaminants. The model allows users to define their own land-use exposure scenarios. The ASTM has developed a methodology for risk-based corrective action.[40] The model allows assessment of risk from various exposure pathways, including groundwater exposure, ingestion of and dermal contact with soil, and vapor inhalation. The model includes a database of the physical, chemical, and toxicolog-

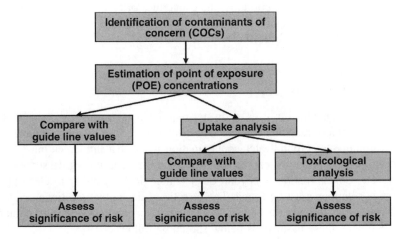

FIGURE 3.15 Basic risk quantification process.

ical properties of a wide range of contaminants. The USEPA has published a comprehensive set of risk assessment tools, including the RAGS (Risk Assessment Guidance for Superfund) guidance documents,[41-43] the IRIS (Integrated Risk Information System) database (available on the Web at www.epa.gov/iris), and the Exposure Factors Handbook,[36] which provides exposure information for a wide variety of mechanisms, including consumption of water, inhalation of vapors, and soil ingestion.

Note that it may not be necessary to quantify the risk from all contaminants of concern. It is possible to use indicator compounds to model groups of contaminants with similar toxicological, chemical, and physical characteristics. For example, benzo-a-pyrene could be used as an indicator compound for the carcinogenic polycyclic aromatic hydrocarbons (PAHs), common contaminants at manufactured gas plant sites.

3.4.3.6 Communication

Effective communication between all stakeholders (site owner, developer, regulators, contractor, consultant, and public) ensures that contaminated sites are managed in the most sustainable way. Risk assessment provides a useful tool for communicating what risks could be present at a contaminated site. Early communication during the risk assessment phase is recommended to ensure that the approach used is acceptable to all parties.

3.5 GROUNDWATER REMEDIATION AND AQUIFER PROTECTION

The decision to remediate is based on a knowledge of the nature and scale of the problems at a site; the risks to the public, water bodies, and other sensitive receptors; the demands of the local regulators; the imperatives of law; the goals of the site owner; and perhaps most importantly, the economics of the situation. Remediation cannot occur if a society cannot afford to conduct the remedial works. In many parts of the world today, this is unfortunately the case.

3.5.1 A RATIONAL APPROACH TO GROUNDWATER CONTAMINATION

The successful remediation of groundwater requires that a number of critical steps be performed before reaching the remedial design stage. The inherent complexities and uncertainties of groundwater contamination mean that implementing groundwater remediation programs can be expensive and time consuming. In some cases, groundwater contamination is beyond our technological capability to clean up, so alternative solutions must be found. These realities dictate that a rational, step-by-step decision-making process be followed. In approaching a groundwater contamination problem, these basic steps below should be followed:

1. *Understand the problems at the site* — Through proper site characterization, a picture of the types, distribution, and concentrations of contaminants and wastes is provided. The characteristics of the groundwater regime are identified. This information serves as the basis for all other activities.

2. *Assess the risks posed by the problem* — Using the tools of risk assessment, in either a qualitative or quantitative fashion, the implications of the problem are determined. Many different types of risk exist (human health, ecological, economic, public relations, personal, and corporate liability), and one or more may be important at the site. Regulatory guidance is available for this step. These risks can be valued and expressed in monetary terms, as is discussed in Part II.

3. *Set remedial goals and constraints for the site* — Once the problem and the risks posed by it are understood, a remedial objective can be set. Understanding the true and total costs and benefits (to all of society) of remediation is a key consideration in setting an appropriate remedial goal and the subject of this book. Constraints that apply to the situation must also be identified. Two of the most important constraints are:

    Cost — What maximum expenditure is warranted to solve the problem? Ideally, costs should be justified by the benefits of remediation.

    Time — How long are stakeholders willing to wait for a satisfactory resolution?

4. *Identify best practicable remedial approach* — Using technical and economic analysis, various possible remedial approaches can be evaluated and compared. A detailed framework for completing economic analysis of remedial approach options is presented in Part III of this book. The approach best able to reach the set goal within the applied constraints is selected, and detailed designs are prepared.

5. *Test and implement the remediation program* — Once a remedial approach has been selected, and with it a preferred technology, the technique should be tested at bench scale (if required) and on a small scale (pilot scale) at site conditions. Based on the results of pilot testing, the system can be scaled up and implemented at full scale.

6. *Monitor results* — Assess remedial progress through careful monitoring, and modify as necessary.

7. *Validate and close* — Confirm and document that remediation has achieved the objectives. If appropriate, close the site.

3.5.2 REMEDIATION TECHNOLOGIES

3.5.2.1 History and Trends

Since the early 1970s, researchers and technology vendors have attempted to develop groundwater remediation technologies that are cost effective and achieve continuously more rigorous clean-up standards. Although groundwater technologies have been available for the last 20 years, their application and effectiveness have varied. In the United States, where strong environmental legislation was enacted in the 1970s, initial clean-up standards were very strict. Considerable effort and money were spent attempting to develop technologies that would meet these mainly numerical standards. Because of U.S. Superfund activities during the 1980s, demand for these technologies was high, and unit costs were high. In some cases, technology vendors made unrealistic claims of performance.

Since then, government-sponsored programs in North America, in Europe, and internationally have promoted the development and critical evaluation of new and more cost-effective remedial technologies. As a result, there are today a number of well established technologies for which the applicability and limitations are well known. Significantly, much of this research has shown that most remedial solutions require the use of more than one, and often several, different technologies. These developments have allowed relatively wide access to the latest remedial technologies and the supporting verified research. However, it is clear from experience and the literature that remedial success requires selecting the right technology for the problem, and then using it properly. Application of technology without knowledge and experience can be a dangerous and expensive proposition.

3.5.2.2 Regulatory Trends

Some of the earliest and most influential regulations dealing with soil and groundwater contamination were formulated in the U.S. in the 1970s and 1980s. This included a major focus on underground storage tanks (USTs). Early U.S. federal and state regulations were based on numerical criteria for specific contaminants. Although these criteria may have provided for an extremely high level of public protection, they were, in many cases, not based on firm toxicological data and were difficult to achieve in practice using existing technologies. More recently, the trend has been away from strict numerical criteria, toward a more risk-based approach (see Section 3.4).

3.5.2.3 Remedial Technology Selection

Once the goals and constraints of remediation have been identified and rationally evaluated, potential remedial approaches and technologies capable of reaching those goals can be evaluated. Those that meet all of the economic criteria and are deemed likely to be able to achieve the technical goal can then be assessed on a comparative basis. In general, the following evaluation process allows sound technical decisions to be made on the feasibility of a remedial approach or technology:

- *Conduct bench-scale or laboratory experiments* — Ideally, these small-scale tests should be carried out under carefully controlled laboratory conditions, using actual media (soil and groundwater collected from the site) and actual contaminants of concern.
- *Conduct prepilot-scale tests* — These tests are conducted in the field at very small scale, usually involving just one well, in the case of an *in situ* (in the ground) remedial method. These tests provide simple yes or no

answers — is this particular technology applicable at this site? If yes, the tests provide design parameters needed to design a pilot-scale test. Figure 3.16 shows a field setup for a prepilot soil vapor extraction test.

- *Conduct a field pilot-scale test of the proposed remediation technology* — This is highly recommended for any *in situ* or on-site remedial method, such as bioremediation, soil vapor extraction, and the like. A pilot test allows techniques, equipment, processes, and materials to be tested under field conditions but on a small scale. Invariably, the findings lead to some modification or adjustment in the final design. Sometimes, the pilot test results are such that the approach is abandoned all together. The rationale for a pilot test is simple — it provides insurance against failure on a large scale. A good rule of thumb is that a pilot test involves about 5 to 10% of the effort, size, and cost of the anticipated full-scale system. At smaller sites, the prepilot and pilot can be combined, or the prepilot scaled up and the pilot dispensed with. At larger, more complex sites, both may be required (Figure 3.17).
- *Final design and optimization of the full-scale system, based on pilot results.*

3.5.2.4 Advancements in Remedial Technologies

In situ technologies for soil and groundwater clean-up have gained wide acceptance in the past several years as cost-effective alternatives to excavation-based remedial methods.[44] In general, *in situ* techniques lead to contaminant control or mass removal

FIGURE 3.16 Prepilot-test setup for soil vapor extraction: well head, knock-out drum, suction blower. Monitoring wells are just outside the photograph to the left and right.

FIGURE 3.17 Monitoring array for a detailed pilot test for soil vapor extraction and bio-venting at a site in Canada. Depending on the scale of the problem, the complexities of the subsurface, and the risks involved, this level of effort and expenditure at pilot scale may be justified.

directly from the subsurface, without resorting to excavation of the material. *In situ* methods involve alteration of subsurface flow, pressure, or chemical or ecological regimes to achieve containment, redirection, removal, or destruction of target contaminants. In general, *in situ* methods are less capital intensive and slower than *ex situ* methods. However, the uncertainty associated with heterogeneous geologic materials means that achievable levels of clean-up are generally lower for *in situ* methods. Other disadvantages include difficulty in confirming a given level of clean-up and the possibility of by-passing pockets of contamination. However, each site must be considered separately and the suitability of each available technique evaluated critically.

Early *in situ* remediation involved the use of interception trenches, installation of low-permeability barriers, or pumping of wells to control contaminant migration and remove contaminants for treatment at surface (pump-and-treat).[45–47] Over the past several years, considerable research effort has led to the development of a wide range of new processes. Many, such as the use of horizontal wells,[48] are simply adaptations of traditional methods, such as pump-and-treat. Semipassive barrier techniques, such as funnel and gate, are enjoying greater application.[49] The use of biological systems to treat a variety of contaminants is also gaining acceptance, including *in situ* bioremediation and bioventing[50,51] and the use of engineered wetlands to treat pumped groundwater contaminants.[52,53] Monitored natural attenuation (MNA) of contaminants has also become a popular approach, allowing the natural retardation and degradation processes occurring within most aquifers to remove contaminants or halt their migration.[54]

3.5.3 EXAMPLE — *IN SITU* REMEDIATION OF HYDROCARBONS USING SOIL VAPOR EXTRACTION

Multiple spills of hydrocarbon liquids from a refinery in South America have resulted in contamination of an underlying sand aquifer.[44] Over the past four decades, hydrocarbon liquids have migrated considerable distances and have now reached a nearby river. Given the depth of the aquifer (about 12 m) and the presence of a large operating refinery complex, *in situ* methods were deemed the only practical way of stabilizing the plume and beginning to remove some of the estimated 100,000 m³ of oil in the subsurface. Piezometers completed as part of the characterization program contained LNAPL thicknesses of up to 3 m, and groundwater surface fluctuations suggested a significant smear zone (Figure 3.18).

FIGURE 3.18 One of the authors holding a bailer containing LNAPL recovered from the groundwater surface at the site.

On this basis, field trials were implemented to test four complementary *in situ* technologies designed to remove hydrocarbons from the aquifer. Specially designed recovery wells were drilled. LNAPL pumping using dual pump systems was applied for removal of the mobile fraction of LNAPL (that which is able to flow to the well). This typically represents only about 10 to 20% of the total hydrocarbon volume.[14] To attack the residual LNAPL in the saturated zone, soil vapor extraction (SVE) was tested. By pumping air from the unsaturated zone, volatile hydrocarbon components are removed.[55] In addition, introduction of fresh, oxygen-rich air into the contaminated zone stimulates biological activity that can degrade hydrocarbons (bioventing).[50] In the saturated zone, high-vacuum multiphase extraction (MPE) was applied (Figure 3.19). This technique uses a strong vacuum to overcome capillary forces that hold oil within the pores of the aquifer, stripping out water and oil simultaneously. In finer-grained materials, this can result in dewatering of the saturated zone around the well screen, allowing vapor extraction to occur.

Pilot test results of these techniques showed promising results. Application of a combination of technologies at this site (SVE, MPE, and liquids pumping) can be expected to remove a significant portion of the most mobile and toxic contamination, reducing the overall risk posed by the contamination.

3.5.4 EXAMPLE — HORIZONTAL WELLS FOR GROUNDWATER REMEDIATION

Horizontal drilling technology has been used for several years in the petroleum industry to achieve enhanced oil recovery, and in the utilities sector for installing

FIGURE 3.19 Pilot multiphase extraction unit at the site, with refinery complex in the background.

lines and cables beneath rivers and other obstacles.[48] It is only recently, however, that horizontal wells have been used for environmental applications. At many contaminated sites where remediation is required, conventional *ex situ* techniques are not practical or cost effective. In some of these cases, horizontal well systems can provide a powerful enhancement to *in situ* remediation methods, such as liquids pumping, soil vapor extraction, and *in situ* bioremediation. In particular, horizontal wells offer distinct advantages in situations where:

- Access to the contaminated subsurface is restricted by the presence of buildings or other structures.
- A relatively thin, flat-lying geological horizon is targeted.
- Distinct vertical geological features, such as fractures, are believed to control contaminant movement and distribution.
- The contaminated zone is of relatively low permeability.

The design of horizontal well systems should be tailored to the specific objectives that the well is expected to achieve. Goals could include hydraulic containment (saturated zone installation with sufficient available drawdown, perpendicular to contaminant plume axis), groundwater contaminant mass removal (for LNAPL, placement just below the groundwater surface and perpendicular to plume axis), or mass removal from the unsaturated zone (placement above the water table to coincide with contaminant occurrence). Design considerations include contaminant type, well placement, orientation of geological features of interest, screen length, approach path, well curvature, location of surface and near-surface obstacles, well materials, and completion strategy. The selection of each of these parameters must be carefully judged to meet the goals of remediation.

At an operational natural gas processing plant, four 140 m horizontal wells have been installed in the unsaturated zone and are being used for soil vapor extraction of hydrocarbon contamination. Average mass removal rates in excess of 1500 kg of hydrocarbon per day per well have been achieved in the first six months of operation. High mass removal rates were partially due to entrainment of free-phase hydrocarbon liquids in the extracted air stream. Cost-effectiveness analysis has shown that the horizontal wells provide greater contaminant mass removal per unit cost than vertical wells at this site. At an abandoned refinery in Canada, a horizontal well has been installed in a thin sand layer to provide hydraulic containment of a plume of free and dissolved-phase hydrocarbons and chlorinated solvents migrating toward a stream. The system is achieving containment of the plume and significant mass removal. Its orientation normal to the direction of outwash channel deposits has provided a more complete capture of migrating contaminants than could be provided by a line of closely spaced vertical wells. These results indicate that, under the right circumstances, horizontal wells can provide a cost-effective alternative to more conventional remedial techniques.

REFERENCES

1. World Bank, *World Development Report: Development and the Environment, World Development Indicators*, Oxford University Press, Oxford, 1992.
2. Freeze, R.A. and Cherry, J., *Groundwater*, Prentice Hall, Englewood Cliffs, NJ, 1979.
3. Blyth, F.G.H. and De Freitas, M.H., *A Geology for Engineers*, 7th ed., Elsevier Science, Amsterdam, 1984.
4. Chow, V.T., Maidment, D.R., and Mays, L.M., *Applied Hydrology*, McGraw-Hill, New York, 1988.
5. Domenico, P.A. and Schwartz, F.W., *Physical and Chemical Hydrogeology*, John Wiley & Sons, New York, 1998.
6. Schwartz, F.W. and Zhang, H., *Fundamentals of Groundwater*, John Wiley & Sons, New York, 2003.
7. Fetter, C.W., *Applied Hydrogeology*, 4th ed., Prentice Hall, New York, 2000.
8. Kruseman, G.P. and deRidder, N.A., *Analysis and Evaluation of Pumping Test Data*, 2nd ed., Int. Inst. for Land Recl. (ILRI), Wageningen, Pub. No. 47, 1991.
9. Cohen, R.M. and Mercer, J.W., *DNAPL Site Investigation*, C.K. Smoley, Boca Raton, FL, 1993.
10. Pankow, J.F. and Cherry, J.A., *Dense Chlorinated Solvents and other DNAPLs in Groundwater*, Waterloo Press, Portland, OR, 1996.
11. Darcy, H.P.G., *Les Fontaines Publiques de la Ville de Dijon*. Victor Dalmont, Paris, 1856.
12. Fetter, C.W., *Contaminant Hydrogeology*, Macmillan, New York, 1992.
13. Schmelling, S.G. and Ross, R.R., *Contaminant Transport in Fractured Media: Models for Decision Makers*, EPA Superfund Issue Paper EPA/540/4-89/004, 1989.
14. API (American Petroleum Institute), *Underground Spill Cleanup Manual*, Am. Pet. Inst., Washington, DC, Pub. No. 1628, 1980.
15. API (American Petroleum Institute), *A Guide to the Assessment and Remediation of Underground Petroleum Releases*, Am. Pet. Inst., Washington, DC, Pub. No. 1628, 2nd ed., 1989.
16. Mackay, D.M. and Cherry, J.A., Ground water contamination: pump-and-treat remediation, *Env. Sci. and Tech,* 23, 6, 1989.
17. USEPA, *Underground Motor Fuel Storage Tanks: A National Survey*, vol. 1., U.S. Government Printing Office, Washington, DC EPA/540/1-86-060, Appendix A-1, 1986.
18. Schmelling, S.G., USEPA groundwater initiative: improvement of pump and treat remediation, *Proceedings 1st GASReP Conference*, Ottawa, January, 1991.
19. USEPA, *Clean-up of Releases from Petroleum USTs: Selected Technologies*, USEPA Office of Underground Storage Tanks, Washington, DC, EPA/530/UST-88/001, 1988.
20. Testa, S.M. and Winegardner, D.L., *Restoration of Petroleum Contaminated Aquifers*, Lewis, Chelsea, MI, 1991.
21. Mendoza, C.A. and McAlary, T.A., Modeling of groundwater contamination caused by organic solvent vapors. *Ground Water*, 28, 2, 199, 1990.
22. Mercer, J.W. and Cohen, R.M., A review of immiscible fluids in the subsurface: properties, models, characterization and remediation, *J. of Contaminant Hydrology*, 6, 107, 1990.
23. Anderson, W.G., Wettability literature survey, part 1. Rock/oil/brine interactions, *J. Pet. Technol.*, Oct., 1125, 1986.
24. Schwille, F., *Dense Chlorinated Solvents in Porous and Fractured Media*, Lewis, Chelsea, MI, 1984.

25. Pinder, G.F. and Abriola, L.M., On the simulation of nonaqueous phase organic compounds in the subsurface, *Water Resour. Invest. Rep.*, 84-4188, 1986.

26. Faust, C.R., Transport of immiscible fluids within and below the unsaturated zone: a numerical model, *Water Resour. Res.*, 21, 4, 48, 1985.

27. Corapcioglu, M.Y. and Baehr, A.L. A compositional multiphase model for groundwater contamination of petroleum products, 1. Theoretical considerations, *Water Resour. Res.*, 23, 1, 191, 1987.

28. Abriola, L.M., *Multiphase Flow and Transport Modeling of the Multiphase Migration of Organic Chemicals: A Review and Assessment*, Electr. Power Res. Inst., Palo Alto, CA, AE-5976, 93, 1988.

29. Abdul, A.S., Kia, S.F., and Gibson, T.L., Limitations of monitoring wells for the detection and quantification of petroleum products in soils and aquifers, *Ground Water Monit. Rev.*, 9, 2, 90, 1989.

30. Kemblowski, M.W. and Chiang, C.Y., Hydrocarbon thickness fluctuations in monitoring wells, *Ground Water*, 28, 2, 244, 1990.

31. Lenhard, R.J. and Parker, J.C., Estimation of free hydrocarbon volume from fluid levels in monitoring wells, *Ground Water*, 28, 1, 57, 1990.

32. Farr, A.M., Houghtalen, R.J., and McWhorten, D.B., Volume estimation of light nonaqueous phase liquids in porous media, *Ground Water*, 28, 1, 48, 1990.

33. Hampton, D.R. and Miller, P.D.G., Laboratory investigation of the relationship between actual and apparent product thickness in sands, *Proc. Conf. Petroleum Hydrocarbons and Org. Chemicals in Ground Water: Prevention, Detection, and Restoration*, NGWA, Dublin, OH, p. 157, 1988.

34. de Pastrovich, T.L., Bradat, Y., Barthel, R., Chainelli, A., and Fussel, D.R., *Protection of Groundwater from Oil Pollution*, CONCAWE, The Hague, 1979.

35. Linkov, I. and Palma-Oliviera, J., *Assessment and Management of Environmental Risks*, Kluwer Academic Publishers, Dordrecht, 2001.

36. USEPA, *Risk Assessment Guidelines for Superfund (RAGS)*, vol. 1, part B, U.S. Government Printing Office, Washington, DC, 1991.

37. Defra and EA, *The Contaminated Land Exposure Assessment Model (CLEA): Technical basis and Algorithms*, (CLR10), The Environment Agency, London, March 2002.

38. U.K. Environment Agency, *Methodology for the Derivation of Remedial Targets for Soil and Groundwater to Protect Water Resources*, R&D Publication 20 (P20), Bristol, 1999.

39. McDonald, M.G. and Harbaugh, A.W., A modular three-dimensional finite difference ground-water flow model, *U.S. Geological Survey Techniques of Water Resources Investigations*, book 6, Ch A1, 1988.

40. ASTM, *Standard Guide for Risk-Based Corrective Action*, ASTM PS-104, 1998.

41. USEPA, *Risk Assessment Guidelines for Superfund (RAGS)*, vol. 1, part A, U.S. Government Printing Office, Washington DC, 1989.

42. USEPA, *Risk Assessment Guidelines for Superfund (RAGS)*, vol. 1, part D, U.S. Government Printing Office, Washington, DC, 2001.

43. USEPA, *Risk Assessment Guidelines for Superfund (RAGS)*, vol. 1, part E, U.S. Government Printing Office, Washington, DC, 2002.

44. Gossen, R.G., Hardisty, P.E., Benoit, J.R., Dabbs, D.L., and Dabrowski, T.L., Site remediation technology advances, *Proc. 15th World Petroleum Congress*, John Wiley & Sons, New York, 1997.

45. Mercer, J.M., Skipp, D.C., and Giffin, D., *Basics of Pump-and-Treat Ground-Water Remediation Technology*, EPA-600/8-90/003 report. Robert S. Kerr Environmental Research, Washington, DC, 1990.
46. CPA (Canadian Petroleum Association), *Effectiveness of Subsurface Treatment Technologies at Alberta Sour Gas Plants: Phase I. Assessment of Subsurface Contamination and Remediation at ASGPs,* Report by Piteau Engineering Ltd., Calgary, 1990.
47. CPA (Canadian Petroleum Association), *Effectiveness of Subsurface Treatment Technologies at Alberta Sour Gas Plants: Phase II. Subsurface Treatment Technologies for Alberta Sour Gas Plants*, Report by Piteau Engineering Ltd., Calgary, 1991.
48. Kaback, D. Horizontal wells, *Workshop Proceedings*, NGWA Outdoor Action Conference, Las Vegas, 1994.
49. Starr, R.C. and Cherry, J.A., *In situ* remediation of contaminated ground water: the funnel and gate system, *Ground Water*, 32, 465, 1994.
50. Moore, B.J., Armstrong, J.E., Barker, J., and Hardisty, P.E., Effects of flowrate and temperature during bioventing in cold climates, in *In-Situ Aeration: Air-Sparging, Bioventing, and Related Remediation Processes*, Hinchee, R., Miller, R., and Johnson P., Eds., Batelle Press, Columbus, OH, 307, 1995.
51. Armstrong, J.E., Moore, B.J., Hardisty, P.E., Khunel, V., and Stepan, D., A field study of soil vapour extraction and bioventing, *Proc. 4th Annu. Conf. on Groundwater and Soil Remediation*, GASReP, Calgary, 1994.
52. Kadlec, R.H. and Knight, R.L., *Treatment Wetlands*, Lewis, Chelsea, MI, 1996.
53. Hardisty, P.E., Currie, L., Murphy, M., Moore, B.J., and Callow, L. Investigation of hydrocarbon attenuation mechanisms within a boreal wetland. *Proc. 5th Annu. Conf. Groundwater and Soil Remediation*, Toronto, 1994.
54. Wiedemeir, S., Rifai, H.S., Wilson, J.T., and Newell, C., *Natural Attenuation of Fuels and Chlorinated Solvents in the Subsurface*, John Wiley & Sons, New York, 1999.
55. Armstrong, J.E., Frind, E.O., and McClellan, R.D., Nonequilibrium mass transfer between the vapor, aqueous and solid phases in unsaturated soils during vapor extraction, *Water Resour. Res.*, 32, 5, 1993.

4 Economic Value of Groundwater — An Introduction

This chapter provides a brief overview of some of the more important and relevant economic principles as they apply to the issue of groundwater contamination and remediation. A considerable body of literature exists on environmental and ecological economics, and interested readers are urged to refer to Pearce and Warford, Costanza, Pearce, Winpenny, Tietenberg, and Perman et al.[1-6]

4.1 ECONOMIC VALUE OF GROUNDWATER — BASIC CONCEPTS

Pearce and Warford state that "the world economy is inextricably linked to the environment because societies must extract, process, and consume natural resources."[1] In fact, we use environmental resources not only as an input to our production process, as this statement implies, but also as a sink for the wastes of that process and as a provider of amenity.

Groundwater, as a renewable resource, fulfills all these roles and contributes to the overall welfare of the society in three main ways:

- As a resource in its own right (through abstraction of groundwater for domestic, agricultural, or industrial use).
- As a contributor to surface water resources, which are used to generate economic value (abstraction, recreation, navigation, etc.).
- As a key part of the hydrologic cycle, contributing to the existence of ecosystems and natural beauty. As well as a resource, groundwater can also be a carrier of contaminants or constitute a risk pathway. For example, in the case of shallow groundwater of poor quality acting as a risk pathway for a sensitive receptor, the value of the groundwater itself may be negligible, but the value of the affected receptor high.

Thus, if groundwater is contaminated, part of the resource base is damaged or, if the pollution is irreversible, eliminated entirely. The potential economic impacts of this "carrier" function of groundwater have not played a crucial role in decision making until recently. One reason for this omission is that the economic (or financial) decision-making systems have only been concerned with the first role of groundwater: as a resource in its own right or merely as an input to production processes.

65

Even this role has not been appreciated well when water abstraction was free of charge or even subsidized. When markets in groundwater abstraction were established (e.g., by abstraction permit system or exchange of water for money in actual markets), the price reflected the importance of groundwater as an input to production and possibly a premium for its abundance or scarcity, depending on the realities of the location or the market.

As with many environmental resources, therefore, the market price of groundwater, when it exists, reflects its value as an input to production. It does not reflect the value of groundwater to society as a contributor to recreational use of surface water, to the sustainability of ecosystems, and to natural beauty. This failure of the pricing system to reflect all services, benefits, or value of environmental resources is referred to as *market failure* in economic literature. In addition, those services or benefits or values that are not reflected in the actual price (or, more generally, all impacts that are not compensated) are termed *externalities*.

Sometimes the reason market prices do not reflect the value of all the services of a resource is that interventions by government policy, such as subsidies (e.g., for irrigation), effectively set a price that is lower than would have been set in free markets. In economic literature, this is known as *policy failure*. Whether the reason is market or policy failure, lower prices for water lead to over consumption and pollution.

4.1.1 TOTAL ECONOMIC VALUE

Environmental economics is concerned with correcting these two types of failures by enabling the market prices to reflect the full set of services and benefits or, as it is called in the literature, the *total economic value* (TEV) of environmental resources. In order to do this, we must first understand and quantify the total economic value of resources as much as possible, and then use this information in decision making. This book is concerned with how to define and quantify the total economic value of groundwater and how to use this information in the decision making about prevention and remediation of groundwater contamination.

Total economic value is the sum of the values placed by individuals on the services (or benefits) provided by a resource. The concept helps us to define the factors or motives behind people's preferences for environmental resources. The total economic value of groundwater (and indeed of all environmental resources) is the sum of use and nonuse values, where (also see Figure 4.1):

- Use values consist of:
  - Direct use values — Uses of groundwater that lead to activities that would not take place if that quantity and quality of water did not exist, including domestic, industrial, and commercial water use, and irrigation for the production of crops or animal feed (groundwater as a resource in its own right).
  - Indirect use values — The value of the contribution made by groundwater to tourism, recreation, and support of the natural ecosystem,

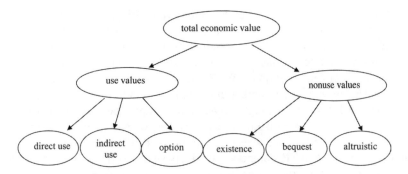

FIGURE 4.1 Total economic value.

    including contributions to the hydrologic cycle (discharge to lakes, rivers, streams, wetlands, and other important surface water features).
- Option values — The premium that certain users may be willing to pay to secure access to and use of groundwater at some time in the future.
- Nonuse values consist of:
  - Altruistic value — Individuals may be willing to protect groundwater or a particular aquifer for the benefit of others who make use of it.
  - Existence value — Individuals may derive value simply from the knowledge that uncontaminated, pristine groundwater, and the ecosystems and hydrological cycle to which it contributes, exists, irrespective of whether the groundwater will ever be used for human consumption.
  - Bequest value — Individuals may affix a certain value to groundwater because of their desire to pass on the resource to future generations.

4.1.2 Measuring Total Economic Value

Given the opportunity, people express their preferences for a resource (or the total economic value they place on it) in two ways:

- They may be willing to pay to maintain a resource as it is in order to secure an environmental improvement (or a benefit) or to avoid environmental degradation (or a cost). When used as a measure of total economic value, this is called *willingness to pay* (WTP).
- They may be willing to accept (monetary) compensation to forgo environmental improvement (a benefit) or to tolerate environmental damage (a cost). When used as a measure of total economic value, this is called *willingness to accept compensation* (WTA).

The advantage of using these measures is the ability to express TEV in units of money: a familiar unit to all and one that allows comparison of the benefits provided by a resource to costs of its protection or alternative uses. In the context of this book, expressing the total economic value of groundwater in monetary terms allows the benefits of remediation to be compared with its financial costs.

The concept of TEV and the preceding measures are based on the concept of price setting in an actual market. In actual (free) markets, the equilibrium price represents the buyers' maximum WTP and sellers' minimum WTA. Individuals with a WTP lower than the market price will not purchase the good in question. Those with WTP equal to or higher than price will purchase the good. The excess of WTP over market price is known as *consumer surplus*. This is the net benefit an individual receives from the consumption of a particular commodity. For a market good, total WTP comprises total consumer surplus (over all units of consumption) plus the total price paid (i.e., total expenditure).

Therefore, when markets in environmental resources exist, data about consumer behavior and prices are used to assess the economic value of that resource. However, this is usually insufficient because of market failure:

- Actual markets do not exist for all environmental resources.
- Even when they do, (free market) prices only reflect the direct use value provided by a resource.

In other words, to use the terminology introduced earlier, when markets fail to capture the full TEV of environmental resources and that of *nonmarket*, and hence unpriced, environmental goods, WTP is wholly composed of consumer surplus (because the price is zero).

In order to make the best use of the different types of the existing evidence, economists use a number of techniques, collectively referred to as *economic valuation techniques*, to quantify the total economic value. These can be examined in four categories:

- Actual market value
- Market price proxies
- Surrogate market techniques
- Hypothetical market techniques

Figure 4.2 shows these four main groups of economic valuation techniques, which component of total economic value they can measure, and how they link with each other.

Before any of the economic valuation techniques can be used, the environmental resource and the changes in its quality and quantity must be defined. Figure 4.2 shows this step of economic valuation analysis as *dose–response relationship*. Dose–response relationships establish links between the change of concern or dose (e.g., pollution or resource degradation) and the impacts of this change or response (e.g., impacts on human health, flora and fauna, and landscapes). In the context of groundwater remediation, dose is the type, magnitude, and route of contamination, and response is the impacts of this contamination on the affected (or potentially affected) receptors, which include human health, the environment, and various uses of groundwater.

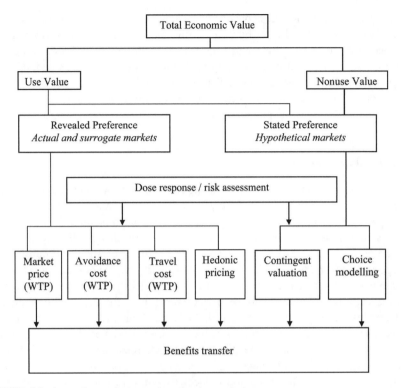

FIGURE 4.2 A typology of economic valuation techniques.

In cases where there is sufficient information and evidence, dose–response relationships can generate a dose–response function and a dose–response coefficient, which quantify the relationship between a unit of contamination and the unit of impact. In other cases, the impacts are described either quantitatively (but with more uncertain links between contamination and impacts) or qualitatively. The risk assessment in the context of groundwater contamination tries to establish such links by following the source–pathway–risk axis.

Figure 4.2 also shows *benefits transfer* as an approach that combines all economic valuation techniques. This approach is used when original economic valuation techniques cannot be applied on a site-specific basis due to limited time and resources available for economic analysis. Further detail on benefits transfer is provided at the end of this section, after the following overview of valuation techniques. As this overview shows, a great deal of interdisciplinary work is required for estimating the economic cost of groundwater contamination.

4.1.2.1 Actual Market Value

The actual market value technique analyzes the price data from actual markets in which the resource of interest is traded. The market price data used to estimate the economic value of groundwater include potable water, industrial water, irrigation water, and value of land sold after the land and aquifer underneath it are cleaned sufficiently for a sale. As these data sources show, this technique can only estimate the use values of groundwater by either using its own price in its various uses or the price of its substitutes for a particular use (e.g., bottled water for drinking water supply). The prices used in such analysis should be net of all taxes and subsidies, which are simply transfer payments. Transfer payments refer to changes that benefit some groups (in the case of taxes, the government) and directly cause a loss to other groups (again for taxes, taxpayers) and hence constitute no net change in society's wealth or welfare.

4.1.2.2 Market Price Proxies

The most common market price proxy is the calculation of what are interchangeably termed *avoidance costs* or *avertive expenditures*. The technique infers benefits of good-quality environmental resources by measuring the consumption of goods and services that substitute for the environmental quality change. Examples include spending on insulation against noise, on water storage or abstraction against lack of supply, and, most relevant in this context, on water filters and bottled water* against contamination (or risk thereof). This inference is clearly not perfect, because the avertive actions are unlikely to be perfect substitutes for the pollution impacts. However (especially in the absence of better data about TEV), avoidance cost can be taken as a minimum proxy for use values (if use values were less than avoidance costs, people would not undertake this spending). On the other hand, avertive expenditures also provide benefits other than the prevention of the environmental problem of concern. For example, insulation reduces exposure to noise and also improves energy efficiency. Similarly, spending on water filters and bottled water could improve the taste of water and prevent exposure to contamination.** Thus, it can be difficult to isolate different types of benefits.

If groundwater contamination can be shown to cause negative impacts on human health, *cost of illness* is another valuation approach that can be used to estimate the economic cost of this impact. The cost of illness calculated here would be the sum of medical costs and the economic value of loss of work days. In order to link the cost of a single illness to the cost of contamination, dose–response relationship is used to estimate the total number of illness cases caused by a given case of contamination. However, due to the difficulties associated with estimating this health dose–response relationship, this method is not widely used in the context of groundwater contamination.

If water is used as an input to a productive process, the method referred to as *productivity change* can also be used. For example, in the case of agriculture, the crop value is determined by a number of factors, including crop type, soil productivity and amount, and quality of water input. The change in crop yield (in physical units), while all else is assumed to remain constant, that is due to the change in the

---

* Bottled water can also be seen as a "substitute" for public water supply provided by aquifers.
** Another version of this technique is *replacement or clean-up costs*. These costs are also inadequate, for the same reason as avoidance costs. Unless there is an (explicit or implicit) social agreement that the environmental resource should be brought back to baseline (predamage) status, clean-up may not be justified. In any case, cost–benefit analysis cannot be used if both costs and benefits are expressed using replacement or clean-up costs, which would always result in a benefit–cost ratio of 1.

quantity and quality of irrigation water is the environmental impact of scarcity or contamination. The economic value of this impact is the amount of marginal crop yield due solely to irrigation multiplied by the market value (net of transfer payments such as taxes and subsidies) of that crop. As with the cost of illness method, this approach is not widely used because of the difficulties in linking the availability and quality of water to the resulting output, such as crop yield.

4.1.2.3 Surrogate Market Techniques

Surrogate market techniques investigate markets that do not trade the environment resource itself but are influenced by it. These techniques are also known as *revealed preference techniques* because people's preferences about environmental resources are revealed through their behavior in surrogate markets. The surrogate market (or revealed preference) techniques can be used only for a subset of environmental effects: those that are reflected in actual markets. Therefore, they can estimate that portion of TEV that is revealed in actual market transactions (or price), that is, use values.

Hedonic pricing technique uses two surrogate markets to estimate the economic value of resources and changes that are not reflected in actual markets. The first such market is housing, which could be used for a number of environmental resources. Hedonic property pricing estimates the price premium that access to good-quality environmental resources adds to property prices. Examples include running water, clean air, beautiful views, and peace and quiet. Time series data are used to examine how property prices change in relation to at least three types of factors:

- Structural (size and type of housing, size of garden, etc.)
- Neighborhood (socio-economic characteristics, public services, etc.)
- Environmental (distance to landfills, industry, access to good-quality water and sanitation services, etc.)

The second surrogate market used by the hedonic pricing technique is the labor market. Supply, demand, and other factors in the labor market determine the price of labor (i.e., salaries and wages). One of these other factors is the risk posed by the occupation on human health, which could be determined by the characteristics of the occupation and the worker, as well as the environmental risk factors. As with the housing market, hedonic wage technique analyzes how wage rates change in relation to factors that are inherent to the occupation, socio-economic characteristics of the workers (e.g., education, experience, age, etc.), and the environmental risks (e.g., risk of using dangerous materials or working in a contaminated environmental when contamination is not known with certainty). In most economic appraisals, it is assumed that the environmental impacts on workers, say, those cleaning a con-taminated site, are internalized by the risk premium such workers receive in their wages. This assumption clearly cannot be made if the existing contamination affects the workers using the site without their knowing about the contamination.

Another source of surrogate market data is travel cost. In developed countries, the *travel cost method* is generally used to assess recreational travel behavior. In that form, travel costs can be used to infer people's preferences for an environmental resource (or its amenity services), based on their willingness to incur costs in traveling to and from a particular site generally for recreation. These costs comprise direct travel costs (e.g., petrol, out-of-pocket expenses, wear and tear, and travel fares) and the value of time. The value, or the opportunity cost, of time is the income that could be generated if time were used not for the activity of concern but for work. Although there is no set rule about this, in practice many studies use 30% of average hourly wage to value leisure time.

In developing countries, however, the same method is used for collection of natural resources, especially water. The travel costs in this context are almost entirely made up of the value of time measured as equivalent to hourly wage rate. In both cases, travel cost data are regressed against the characteristics of the environmental resources, the travel, and the individuals to explain the economic value of the resource.

4.1.2.4 Hypothetical Market Techniques

Hypothetical market techniques elicit individuals' WTP for a specific outcome by way of structured questionnaires. The questionnaires construct hypothetical markets to establish how much people would be willing to pay for environmental benefits or to avoid losses. Conversely, people can be asked how much they would be willing to accept in terms of compensation to forgo environmental benefits or to suffer losses. Because the questionnaires give respondents the chance to state their preferences about environmental resources, these techniques are also known as *stated preference techniques*. The techniques can potentially be implemented for all environmental effects and can also investigate all different motivations behind individuals' WTP (or components of TEV).

There are two groups of stated preference techniques: contingent valuation and choice modeling. Although there are differences between the two, in short, WTP (or WTA) responses are regressed against factors that are thought to influence this response, such as the socio-economic characteristics of the respondents and relevant aspects of the environmental resource.

Contingent valuation questionnaires present, say, an aquifer as a bundle of attributes. These attributes can be quality or quantity of drinking water supply, quality of surface waters charged by the aquifer, and so on. The attributes can take both positive and negative values, depending on how they are affected by the changes to the aquifer. The technique generates estimates of WTP or WTA for the entirety of this bundle. The WTP/WTA question can be asked:

- In an open-ended format: "What are you willing to pay for …?"
- In a dichotomous choice format: "Are you willing to pay $X?"
- By using a payment card design, asking respondents to choose the maximum amount they would be willing to pay from the amounts listed on the card

Also asked are questions about attitudes toward the subject of the questionnaire (here contamination of groundwater) and socio-economic characteristics of respondents. Therefore, the technique creates interesting information beyond WTP or WTA, information that could be used in negotiating remediation objectives and approaches.

If the objective is to estimate how much people value individual attributes of an aquifer, then *choice modeling* is the preferred approach. The technique asks the respondents to choose between different aquifer remediation schemes that impact the different attributes differently. One of the attributes is always cost or price, which enables the analysis to infer WTP/WTA from the choices people make. Table 4.1 provides an overview of economic valuation techniques and their coverage of TEV.

4.1.2.5 Using Economic Value Estimates–Benefits Transfer

In practice, it is often not possible to undertake original valuation exercises to estimate the total economic value of a resource every time this information is needed. Similar to any other analysis, economic valuation takes time and could be limited by data availability. Whether or not an original valuation study can take place for the analysis of a given groundwater contamination case depends on:

- Time and effort required for economic valuation in relation to the overall cost of remediation effort and, possibly more importantly, the cost of a wrong decision*
- Time and effort required for economic valuation in relation to critical importance of the environmental resource and the environmental or human health impact
- Type of economic value (use, nonuse, or both) to be estimated
- Availability of data that may require undertaking environmental impact analysis, risk assessment, and other methodologies that generate the necessary impact data

Instead of original economic valuation studies, what is often done for individual project appraisal studies is *benefits transfer*. Benefits transfer is an approach to adjusting and adopting the estimates of values from previous studies to be used in the analysis of the costs and benefits of the project in question. Time and effort saved by applying benefits transfer instead of original economic valuation studies serve two purposes. First — and this is the more important reason — benefits transfer can be used to get an indication of the order of magnitude of an environmental impact. This helps to determine whether an impact is significant (i.e., when estimates have a high monetary value and when the situation is worthy of further investigation by an original study). Second, benefits transfer may be suitable for direct use in cost–benefit analysis (CBA) (see Chapter 5).

In implementing benefits transfer, the three most common procedures are:

- To transfer an average WTP estimate from an original valuation study
- To transfer WTP estimates from meta-analyses**
- To transfer a WTP function

---

* Nothing is "too expensive" in isolation. Expense is relative to the remedial budget, likely cost of environmental damage, and the cost of making a mistake (remediating too little or too much).

** *Meta-analysis* is typically defined as the statistical analysis of the results of a number of empirical studies. It is an attempt to derive valid generalizations by identifying consistency of results across different studies.[9]

TABLE 4.1
Overview of Economic Valuation Techniques

Techniques	General Application	Advantages	Disadvantages
		Actual Market Value	
Market prices	• Environmental resources that are traded in actual markets	• Modest economic data requirements	• Estimates use value only • Markets do not exist for most environmental resources
		Market Price Proxies	
Avertive expenditures Cost of illness Productivity change	• Water quality, noise nuisance, air pollution and radon contamination (potentially all known and available environmental problems — limited by availability of markets)	• Modest economic data requirements • Use of real market data	• Estimates use value only • Problems arise when individuals make multiple avertive expenditures, there are secondary benefits of an avertive expenditure, and avertive behavior is not a continuous decision but a discrete one (e.g., double glazing is either purchased or not) • Cannot cover environmental changes without precedence • Could be difficult to establish the link between contamination and human health or environment in physical quantities

Surrogate Market (Revealed Preference)

Method	Description	Advantages	Disadvantages
Hedonic property pricing	Premium added to the property prices due to favorable environmental and neighborhood charateristics of the surrounding area (e.g., air quality noise, landscape, water supply, etc.)	• Use of real market sdata	• Estimates use values only (i.e., underestimates TEV) • May have substantial data requirements • Cannot cover environmental changes without precedence
Travel cost method	Total recreational value associated with national parks, reservoirs, woodlands, forests, wetlands, coastal zones, fisheries, etc.		

Hypothetical Market (Stated Preference)

Method	Description	Advantages	Disadvantages
Contingent valuation (CV)	• Potentially can be applied for all environmental resources and changes, even though the application is limited by the complexity of the questionnaire	• Can estimate use and nonuse values • Estimates WTP per attribute (CM) • Does not directly ask WTP questions (CM) • Can cover all environmental changes with or without precedence • Completed surveys give full profile of affected population • Not limited by readily available data because questionnaire collects data	• Generally more expensive than the other two techniques • CM is not yet as widely tested as CV • Some versions of CM are not based on economic theory • Hypothetical nature of the exercise is sometimes criticized
Choice modeling (CM)			

If there is an *a priori* reason to expect a difference between the original study information and the new project context, WTP values (and function coefficients) may be adjusted to reflect this expectation.

For a benefit transfer exercise to be "valid," certain conditions should be met.[7,8] These conditions are widely recognized to be the following:

- Studies included in the analysis must be sound.
- Studies should include a WTP bid function (regression analysis showing the influence of explanatory variables on WTP).
- *Study site* (the location of the original study) and the *project site* (the new valuation context) must be similar in terms of population and site characteristics, or differences in characteristics must be adjusted.
- Change in the provision of the good being valued at two sites should be similar.
- Property rights should be the same across the sites.

These conditions may also serve as health warnings that recognize the limitation of the benefits transfer approach. Specifically, individuals at two different locations may value environmental resources differently due to differences in:

- Socio-economic characteristics
- Physical characteristics of a particular site
- Proposed changes in the environmental goods at each site
- Availability of substitutes for a particular site[10]

Moreover, the opportunities to experience environmental benefit between two sets of individuals may be different.[11] Hence, WTP in the original study site may not provide an accurate estimate of WTP in the new site.

The next section gives examples of the economic valuation studies that investigate the economic value of groundwater and hence the benefits of preventing and remediating contamination. To evaluate the applicability of valuation studies for the economic analysis of groundwater remediation, a number of study screening or selection criteria are implemented. The following list briefly outlines these:

- Study subject — A given environmental impact (e.g., degraded water quality) could come about due to a number of different causes (e.g., agricultural, industrial, and domestic pollution sources). Individuals' preferences for avoiding this impact may be influenced by the cause of the impact, so before studies are used the relevance of the source of the impact and subject of the study should be assessed.
- Study context — It is important to match the valuation context in an original study to the context of economic analysis. In the context of measuring nonmarket benefits, studies that elicit WTP for an improvement consider a *compensating surplus*, whereas studies that elicit WTP

to avoid a loss or to maintain the status quo consider an *equivalent loss*.[12,13]*

- In economic valuation, the valuation context should determine the appropriate measure on the basis of the present level of the resource in question and the envisaged final level of the good after the change in provision. For benefits transfer exercises, the context similarity between study and project sites should be maintained. However, in some instances it is necessary to consider a different context than that of the site in order to apply economic values.
- Study origin — Typically, it is recommended that benefits transfer exercises in a given country should use studies relating to that country. However, it is sometimes inevitable that studies from other countries with similar socio-economic characteristics are considered. This is particularly the case when significant gaps exist in the valuation literature for a particular country.
- Study methodology — Studies should be grounded in economic theory and use robust valuation methodologies. In addition, studies considered for benefits transfer should demonstrate that valuations given can be assessed for their reliability and validity by providing statistical results from the analysis undertaken.
- Study date — The date of the original study is relevant because the design and implementation of economic valuation techniques have developed over the past 15 years or so.

The discussion so far has concentrated on estimating individuals' WTP (or WTA) for a change in environmental resources. What is needed for economic analysis, however, is the aggregate economic value, and this requires the identification of the *affected population*. Generally, there are three possible choices of affected population:

- Residents in the vicinity of the resource, good, or impact
- Visitors to a location
- Regional, national, and even global population

Typically, residents and visitors are considered users of a resource, whereas the general population includes both users and nonusers and should be considered where nonuse values are deemed significant.

In practical terms, market data or market proxy data are quicker and easier to use in a project-level cost–benefit analysis. Although these data are incomplete, they still constitute an improvement over the current practice in the context of groundwater contamination, in which wider environmental and social benefits of remediation are totally ignored.

---

* The terms *compensating surplus* and *equivalent loss* are special cases of the more general welfare measures of compensating variation and equivalent variation. The former cases arise when assessing changes in nonmarket goods.[14]

4.1.3 Using the Total Economic Value Information in Decision Making

The preceding discussion already alludes to the various uses of quantitative evidence of TEV in decision making. These include:

- Project, program, and policy appraisal
- Setting priorities for environmental policy
- Determining the level of environmental taxes or other economic instruments, such as tradable permits and design of voluntary agreements
- Estimating the amount of environmental liability
- Green national accounting
- Green corporate accounting

For the purposes of this book, the most important item on the preceding list is project, program, and policy appraisal. The most common analysis techniques are cost–benefit analysis and cost-effectiveness analysis.

4.1.3.1 Cost–Benefit Analysis

Cost–benefit analysis (CBA) is a framework for comparing the monetary value of benefits of a project or policy with the monetary value of costs. It can answer the two most important decision-making questions: "Should we remediate the contamination? If so, to what level of quality?" The optimal remediation level, then, is the level at which the net benefit (benefits minus costs) of remediation is maximized.

A *benefit* is defined as anything (financial, environmental, or social) that increases human well-being, and a *cost* as anything that decreases human well-being. The changes are measured against a common baseline (usually the do-nothing or business-as-usual scenario), and benefits and costs are defined under the "with" and "without" remediation scenarios. This also relates to the concept of opportunity cost, which implies that resources should be valued in terms of the benefits they would generate not for the purpose in context (with) but in their next-best use (without). In the context of groundwater remediation, benefits are the environmental damage avoided plus other benefits of clean-up. Costs, on the other hand, consist of financial and environmental costs of undertaking remediation.

In this context, the factor affecting human well-being is the risk of groundwater contamination and its related impacts on human health and ecosystems (identified as components of the TEV of groundwater). In turn, human well-being is determined by whatever people prefer and quantified using the concepts of WTP and WTA, whether from market data or nonmarket economic valuation techniques.

4.1.3.2 Cost-Effectiveness Analysis

Cost-effectiveness analysis (CEA) differs from CBA insofar as benefits of remediation are usually not quantified. It requires that a level of remediation for a given site is agreed for regulatory, political, or social reasons. CEA then becomes a framework to establish the least costly method to achieve this given level of remediation. CEA can answer the question of how to remediate, but it cannot answer the crucial question of whether the level of remediation (or remediation objective) given is the socially optimal level. In the groundwater remediation literature, the term cost–benefit analysis is generally used for what is in fact cost-effectiveness analysis, where benefits refer to advantages of different remedial technologies.

There are at least three reasons why undertaking a fully quantitative CBA or CEA may not be possible:

- Underlying physical data about contamination do not exist. If, for example, no one has carried out a risk assessment of, say, a given chemical, it will not be possible to estimate the economic benefit of reducing that chemical in the environment.
- Underlying physical data may exist but not in a form suitable for monetary expression. Recall that monetary values reflect preferences. Now, suppose the physical data take the form of "reduction of X tons of chemical oxygen demand (COD)" in an aquifer. Individuals do not have measurable preferences for COD. What they prefer is more or less groundwater quality. The object of preferences does not correspond to the physical measure of the environmental change. This is the so-called correspondence problem, which can potentially be addressed by expressing the environmental effects in units of impacts for which individuals have preferences (e.g., effect on taste or the visual appearance of water due to change in COD levels).
- Relevant physical data may exist and may correspond to what people value, but the economic research may not have been done. Consider biological diversity. There are numerous studies of willingness to pay to conserve biological resources (e.g., endangered species and habitats) but hardly any that tell us what people's preferences are for diversity *per se*.

4.1.3.3 Multicriteria Analysis

When costs and benefits are expressed by a mix of monetary and nonmonetary units, a weighting or scoring system that allows for simultaneous comparison of all costs and benefits is needed. One way of doing this is to undertake a multicriteria analysis (MCA). MCA is a two-stage procedure. The first stage identifies a set of goals or objectives (e.g., different levels of remediation) and then seeks to identify the trade-offs between the alternative objectives or between alternative ways of achieving a given objective (e.g., different engineering alternatives for remediation). The second stage seeks to identify the best policy by attaching weights to the various objectives. A set of such weights can be the monetary values of financial, environmental, and social costs in question. Therefore, MCA is capable of combining monetary and nonmonetary values for different costs and benefits.

All three tools of analysis require information about the costs and benefits of remediation, whether this is expressed in monetary (e.g., U.S.$) or nonmonetary (e.g., reduction in risk of off-site contamination by 1%) terms. Site investigation and risk assessment (together with identification of engineering alternatives) help to gather information on financial, environmental, and social costs and benefits. More detailed assessment of these economic assessment or appraisal techniques (and others) is presented in Chapter 5, with specific implementation to cases of groundwater remediation in Part IV.

4.2 ECONOMIC VALUE OF GROUNDWATER — REVIEW OF LITERATURE

Section 4.1 briefly reviewed the concept of economic value and how this can be measured in monetary units. This section summarizes a selection of economic valuation studies that use different techniques.

4.2.1 STUDIES USING ACTUAL MARKET VALUE TECHNIQUES

Kulshreshtha estimated a range of direct use values for an aquifer, based on the value to rural households, small nonfarm communities, larger communities, and commercial and industrial users.[15] Table 4.2 presents the values determined by this study. Indirect use (recreational value) was assumed to exist by virtue of groundwater recharge to a major river that runs through a national park within the study area. The river and park support a diverse and valuable ecosystem. The park is a major visitor attraction and contributes significantly to the economy of the area. Indirect use was estimated by the value of the recreational activities supported by the river in the park, expressed as the number of visitor-days to the park (a readily available statistic) and WTP for the park amenity (surveyed at U.S. $3.8 per person per day). The value of the aquifer was also estimated by considering the opportunity cost of groundwater, namely the cost of developing alternative water supplies of similar quantity, quality, and reliability. The costs of various alternatives were estimated, including accessing a nearby lake by pipeline ($1885/Ml).

TABLE 4.2
Summary of Average Water Values for Different Uses in Carberry Aquifer Region in 1990

Type of Use	Average Value of Water (U.S. $/Ml)
Irrigation	355
Other farm use	455
Domestic	87–407
Industrial	12–24
Commercial	69–282
Defense (military base)	131–696

Source: From Kulshreshtha, S.N., *Social Science Series, 29*, Environmental Conservation Service, Environment Canada, 1994.

Kulshreshtha suggests that the value of a given aquifer reflects the "economic welfare of people living in the region served by the aquifer.[15] If the aquifer were not present, or was destroyed, the economic welfare of society would diminish by this amount, on an annual basis, for the remaining productive life of the aquifer." Thus, if we accept the author's definition of an aquifer as a renewable, long-term resource, it can be seen that the economic implications of irreversible aquifer damage could be considerable. The paper also brings forward the notion that an aquifer may have additional value in a regional development context. The economic activity generated by virtue of the existence and use of the aquifer at a local scale may create a multiplier effect in the region.

4.2.2 STUDIES USING SURROGATE MARKET VALUE TECHNIQUES

Government-run water supply systems have been the focus of some avoidance cost (AC) studies. For example, Nielson and Lee calculated annual pesticide removal costs from groundwater at between $333 and 67 per household for water supplies serving 5000 and 500,000 customers, respectively.[16] Household AC studies from various parts of the U.S. are discussed.

Abdalla reviewed five ACM studies that attempted to measure household costs resulting from groundwater contamination.[17] Typical actions taken by households included purchasing bottled water and installing water filtration systems. Abdalla examined a community in Pennsylvania served by a public water supply that was contaminated with organic chemicals.[18] Of all households, 96% were aware of the contamination, and 76% were undertaking their own averting actions. Costs averaged $252 per year for each household choosing to avoid the contamination.

Rural communities in Virginia served by private groundwater wells were studied by Collins and Steinback.[19] They found that 85% of households informed about groundwater contamination engaged in some form of averting action, including hauling water and end-of-pipe treatment. Weighted average economic avoidance costs were estimated at $1090 per household for organic contamination problems. The study concludes that economic avoidance costs, as a measure of the benefit of groundwater protection, are highly dependent on local conditions and the knowledge that a problem exists. However, it is clear that avoidance actions of households can be significant in economic terms.

These examples of AC studies are for residential properties; little published information exists for the commercial sector. Hardisty uses the avoidance cost method to determine the private (internal) benefits of remediation.[20] In the case of a firm or problem holder contemplating remediation, avoidance costs can be seen as potential benefits of going ahead with remediation. These include avoiding the risk of litigation (and the considerable costs that may be involved), avoiding fines, averting public relations damage that could result in loss of sales revenue, and preventing control orders or shutdowns that may result in lost production and revenue.

4.2.3 STUDIES USING HYPOTHETICAL MARKET VALUE TECHNIQUES

Edwards, in one of the first studies of its kind, studied the WTP to protect a water supply aquifer in Cape Cod, Massachusetts, from nitrate contamination.[21] His survey determined a WTP of $1623 per household per year but also found that the uncertainty associated with valuation was so great that he could form no clear conclusions from the results.

Hanley used CV to show that individuals in the U.K. had a WTP of about $23 per person per year to guarantee that water supplies meet nitrate standards.[22] He commented that Edwards's WTP value was much too high. We must assume, however, that the WTP for protection from other, more toxic forms of contamination would be at least as much as for nitrate contamination.*

Powell used contingent valuation in 15 communities — four in Massachusetts, four in New York, and seven in Pennsylvania — as part of a research project sponsored by the U.S. Geological Survey, U.S. Department of the Interior.[26] All the towns chosen rely on groundwater for their drinking water, and seven had had contamination by either trichloroethylene or diesel fuel. The sources of these stressors were agricultural fertilizers, toxic chemicals, landfills, accidental spills, underground storage tanks, and septic tanks. The study population was familiar with contamination of the public water supply. The baseline condition was the prevailing groundwater quality at the time with potential contamination sources present.

The change modeled was an areawide special water protection district that would develop and implement pollution prevention policies designed to suit the needs of the community. Questionnaires, mailed to a total of just over 2100 households in 15 towns, returned an overall response rate of about 50% (n = 1041). Among other questions, respondents were asked their WTP for additional water bill charges for public water users and in the form of an addition to property tax for private well users.

Mean annual household willingness to pay for increased water supply protection (in 1989 U.S. dollars) was $62. Experience of a contamination incident was found to increase mean annual WTP per household by $26. Willingness to pay per household per year was $26 more for those who perceived contamination in the near future as likely, compared to those who considered contamination unlikely. Respondents on public supply were more concerned about potential contamination sources than those with private wells; nevertheless, those with private wells were willing to pay $14 more than those on public supply. This last point might be explained by the fact that those with private wells paid nothing for water at the time, whereas those on public supply already paid water bills.

Schultz used contingent valuation to investigate the WTP for protecting groundwater in Dover, New Hampshire.[27] Dover relies on groundwater sources stored in aquifers for its water supply. At the time of the study, surrounding towns with similar water supply sources had recently had groundwater pollution problems due to leaching of chemicals and toxic wastes from underground storage sites. Dover had not yet experienced groundwater pollution problems, except an incident of benzene contamination that resulted in the closing of two city wells. The city of Dover was in the process of drafting a groundwater protection ordinance as economic analysis was

---

* Several other studies are available in the literature dealing with the benefits of protecting groundwater from nonpoint-source pollution from agricultural fertilizer application leading to nitrate contamination.[23–25] These studies are useful, in that they reflect very similar conditions that might be experienced through releases of contaminants from industrial sites into aquifers.

undertaken. The ordinance goal was "to promote public health, safety and general welfare by protecting and preserving the quality of existing and future groundwater supplies from adverse or detrimental land use, development or activities."

A contingent valuation survey was mailed to 600 Dover property owners whose names were chosen from Dover's 1988 property tax list. The WTP was sought for protecting the aquifer against potential contamination. Payment vehicle was specified as an increase in property taxes. Mean WTP value or bid, in 1988 prices, was $215 per household per year. However, the authors claim that the mean WTP value of $129 per property owner per year (in 1988 dollars) found when bids of $500 and above were excluded can be considered the best representative of mean WTP for groundwater protection in Dover.

Bergstorm and Dorfman investigated the WTP of the residents of Daugherty County, Ohio, for ensuring the quality of groundwater to be fit for drinking and cooking.[28] Potential sources of contamination were agricultural pesticides and fertilizers including, but not limited to, Aldicarb, Atrazine, Alachor, Carbofuran, and nitrates. A mail survey was administered to a random sample of Dougherty County citizens (600 returns), chosen from a list of registered voters in the county. Four different versions of the survey, which varied the amount of information about the characteristics of groundwater (including the likelihood of contamination), were distributed to random subsamples. Across the different options, WTP was estimated to be between $320 and $2360 per household (1994 dollars). The higher end of the range represents certain or very high probability of contamination.

Stenger conducted a contingent valuation survey of 100 randomly selected households from each of 10 different municipalities in Alsace, France, between March and June, 1993.[29] The aquifer is highly vulnerable to various sources of pollution, such as those from the intensive use of pesticides and fertilizers in agriculture, waste dumping by industries and municipalities, increased road traffic, accidental pollution of the Rhine aquifer, and the proximity of the groundwater to the surface. Three of the surveyed municipalities had been exposed to water contamination in the past or during the survey period. Under the baseline conditions (of doing nothing), the aquifer would have been entirely contaminated within 10 years or so. WTP was sought for a protection program for the aquifer so that the risk of contamination would be reduced to near zero.

The respondents included users and nonusers of the aquifer. The questionnaire elicited information about the households' water consumption; the respondents' knowledge and opinion about groundwater quality, bottled water, exposure, and pollution and its health effects; willingness to pay for pollution prevention; and socio-economic background. The survey generated a total of 817 questionnaires suitable for analysis. Two different methods of payment (referendum or dichotomous choice [yes/no] and open-ended) and two versions of the questionnaire were used. The two versions presented different levels of reliability of the proposed preservation program. Estimates of the mean WTP were close to the observed mean of $927 per household per year (1993 dollars).

Poe studied CV to estimate a damage function for nitrate exposures based on actual water test results on groundwater supply wells.[30] Damages were estimated as WTP for protecting individual well supplies to a 10 mg/L health-based standard. In a review of the available literature, Poe concludes that "people simply do not have well-informed reference conditions, and thus it is unlikely that values collected under these conditions would reliably predict WTP for a population *actually experiencing* groundwater contamination."

He argues that alternatives that provide respondents with hypothetical exposure scenarios also have limitations. Again, the link to setting groundwater protection *policy* is stressed — economics must play a key role. Poe conducted his own WTP survey for private wells in Wisconsin, using actual groundwater nitrate values measured by kits provided to each respondent.[30] With knowledge of the nitrate levels in their own wells and armed with information describing the health effects of nitrates in drinking water at various concentrations, respondents provided their WTP for groundwater protection. This approach contrasts with previous WTP studies based on hypothetical conditions ("Suppose your tap water were contaminated by nitrates to a level of X ...").

Poe argues, with some success, that when faced with actual conditions, respondents are more likely to provide realistic WTP values. Interestingly, this study of 332 households found a concave relationship between nitrate level and WTP: willingness to pay for protection initially rises quickly at low levels of contamination, and then levels off markedly above the health-based threshold concentration (at about $500 per household per year). The author concludes that this result, featuring a WTP that has an upper bound, is consistent with the opportunities for ready substitutes, such as bottled water.

The issue of taste and odor effects on water supplies has been well documented.[31] Contaminants such as MtBE, for instance, create taste and odor problems in water at concentrations well below current health-based concentration limits.[32] Economic impacts from nontoxic taste and odor effects must also be considered.

A couple of studies by Boyle and Powell reviewed a number of contingent valuation surveys to investigate similarities and differences between them. [33,34]

Boyle and colleagues examined the state of groundwater valuation information available and concluded that, up to 1994, only eight original studies had considered the economic benefits of protecting groundwater quality.[33] They examined these studies statistically and probabilistically and concluded that the data were difficult to use in a systematic way. Definitions of what constituted contamination were inconsistent among the studies. Benefits transfer approaches were difficult to use because problems were inherently too site specific. The researchers found that results of contingent valuation studies were highly dependent on the design of the survey instruments, and thus were generally not accepted by noneconomists. This, in part, reinforces our introductory statement about the gulf between practitioners in the technical and economic fields. However, they concluded that despite these limitations, contingent valuation studies were not producing "random noise," but reflecting fundamental attitudes.

Powell and colleagues investigated the use of contingent valuation information as a tool to persuade decision makers to implement water supply protection policies.[34] They conducted a contingent valuation survey in three northeastern states in the U.S.

of annual household WTP for increased groundwater supply protection. Results show a mean WTP of $61.55 per household per year, but with a relatively high standard deviation (84). Interestingly, they found that knowledge of groundwater issues was not a significant predictor of the survey's outcome. In fact, as expected by economic theory, household WTP could be predicted based on income, experience of a previous contamination incident, and type of water supply (public or private). They confirmed the results of other contingent valuation studies, finding a very low correlation coefficient for the data. Although acknowledging the limitations of contingent valuation surveys, the study concludes that as a technique, economic valuation can indeed persuade decision makers of increased public support for groundwater protection.

4.2.4 SUMMARY

The studies reviewed above are summarized in Table 4.3. The estimates presented in the last column of the table are not directly comparable since economic values are influenced by:

- Characteristics of the aquifer and contamination (e.g., type of pollutant, actual and perceived risk of contamination and availability of substitutes)
- Characteristics of the affected population (e.g., socio-economic characteristics, prior experience of contamination, and private or public water supply).
- Characteristics of the valuation methodology adopted (e.g., type of valuation technique used, year of data, the detail of the information study provides about the contamination [especially in the case of hypothetical market techniques])

The literature reviewed is consistent in the view that the wider benefits of groundwater protection and remediation can be quantified, at least partly. However, it is clear that monetization of the benefits is fraught with difficulty and requires significant effort and expense in its own right. All of the studies reviewed here include statements highlighting the need for more research, particularly into nonuse values (which many felt could be quite significant), valuation techniques themselves, and the need for more case studies and real data.

To conclude this chapter, it is interesting to refer to Abdalla.[17] In the authors' opinion, compartmentalization and specialization of economic studies in the realm of resource valuation means that economists are limiting their ability to help policy makers integrate information about the value of groundwater and the costs of impacts on the resource. Broader, more innovative economic approaches could be overlooked as specialists delve ever deeper into their own subdisciplines. Again, this echoes one of the themes of this book: the need for a rapprochement between economists and technical groups working in this area.

TABLE 4.3:
Summary of Economic Valuation Literature

Author	Country	Valuation technique	Context	Economic value estimate (year of data)
Kulshreshta[15]	Canada	Actual market value	Economic value of aquifer	US$12-696/ML (depending on the use of groundwater) (1990)
Nielsen and Lee[16]	U.S.	Avoidance cost	Investment to remove pesticides from aquifer	US$67-333 / household (depending on the size of the community) (1987)
Abdalla[18]	U.S.	Market price proxy	Purchase of water filters and bottled water	US$252/household/ year on average (1994)
Colin and Steinback[19]	U.S.	Avoidance cost	Hauling water from elsewhere and end-of-pipe treatment	US$1090 / household (1993)
Hanley[22]	U.K.	Hypothetical market	WTP to guarantee that drinking water meets nitrate standards	US$23/person/year (1991)
Powell[26]	U.S.	Hypothetical market	WTP to set up an area-wide fund to address sources of groundwater pollution	US$62-88 /household/year (upper limit estimate is from respondents with prior (or expected) contamination)
Schultz[27]	U.S.	Hypothetical market	WTP to protect aquifer against potential future contamination	US$129/household/ year (1988)
Bergstorm and Dorfman[28]	U.S.	Hypothetical market	WTP to ensure groundwater fit for drinking	US$320-2360/ household (1994) (the lower and upper limit estimates reflect low and high probability of contamination)
Stenger[29]	France	Hypothetical market	WTP for pollution prevention	US$927 / household/ year (1993)
Poe[30]	U.S.	Hypothetical market	WTP to protect groundwater from nitrate pollution (actual measurements in respondents' wells)	US$500 / household/ year (1998)

REFERENCES

1. Pearce, D.W. and Warford, J.J., *World without End*, World Bank, Oxford University Press, Oxford, 1993.
2. Costanza, R., Ed., *Ecological Economics*, Columbia University Press, New York, 1991.
3. Pearce, D.W., *Economic Values and the Natural World*, Earthscan, London, 1993.
4. Winpenny, J., *The Economic Appraisal of Environmental Projects and Policies: A Practical Guide*, OECD, Paris, 1995.
5. Tietenberg, T., *Environmental and Natural Resource Economics*, 4th ed., HarperCollins, New York, 1996.
6. Perman, R., Ma, Y., McGilvray, J., and Common, M., *Natural Resources and Environmental Economics*, 2nd ed., Longman, London, 1999.
7. Boyle, K.J. and Bergstrom, J.C., Benefit transfer studies: myths, pragmatism, and idealism, *Water Resour. Res.,* 28, 3, 657, 1992.
8. Desvousges, W.H., Naughton, M.C., and Parsons, G.R., Benefits transfer: conceptual problems in estimating water quality benefits using existing studies, *Water Resour. Res.,* 28, 3, 675, 1992.
9. Brouwer, R., The validity and reliability of environmental benefits transfer, PhD diss., Univ. of East Anglia, 2000.
10. Bateman, I.J. and Willis, K.G., Eds., *Valuing Environmental Preferences: Theory and Practise of the Contingent Valuation Method in the US, EU and Developing Countries*, Oxford University Press, Oxford, 1999.
11. Bergland, O., Magnussen, K., and Navrud, S., Benefit transfer: testing for accuracy and reliability, in *Comparative Environmental Economic Assessment*, Nijkamp, P. and Willis, K. Eds., Edward Elgar, Cheltenham, U.K., 1995..
12. Bateman, I.J. and Turner, R.K., Valuation of the environment, methods and techniques: the contingent valuation method, in *Sustainable Environmental Economics and Management: Principles and Practice*, Turner R.K., Ed., Belhaven Press, London, 120, 1993.
13. Bateman, I.J., Willis, K.G., and Garrod, G.D., Consistency between contingent valuation estimates: a comparison of two studies of U.K. National Parks, *Regional Studies*, 28, 5, 457, 1994.
14. Hanley, N. and Spash, C., *Cost–Benefit Analysis and the Environment*, Edward Elgar, Cheltenham, U.K., 1993.
15. Kulshreshtha, S.N., *Economic Value of Groundwater in the Assiniboine Delta Aquifer in Manitoba*, Social Science Series, 29, Environmental Conservation Service, Environment Canada, Ottawa, 1994.
16. Nielsen, E.G. and Lee, L.K., *The Magnitude and Costs of Groundwater Contamination from Agricultural Chemicals: A National Perspective*, U.S. Dept. of Agriculture, Ag. Econ. Rep. No. 576, 1987.
17. Abdalla, C.W., Groundwater values from avoidance cost studies: implications for policy and future research, *Amer. J. Agr. Econ.,* 76, 1062, 1994.
18. Abdalla, C.W., Roach, B.A., and Epp, D.J., Measuring economic losses from groundwater contamination: an investigation of household avoidance costs, *Water Res. Bull.,* 26, 451, 1992.
19. Collins, A.R. and Steinback, S., Rural household response to water contamination in West Virginia, *Water Res. Bull.,* 29, 199, 1993.

20. Hardisty, P.E., Bracken, R.A., and Knight, M. The economics of contaminated site remediation: decision making and technology selection, *Geol. Soc. Eng. Geol. Sp Pub.* 14, 63–71, London, 1996.
21. Edwards, S.F., Option prices for ground water protection, *J. Env. Econ. Manage.*, 15, 474, 1988.
22. Hanley, N., The economics of nitrate pollution control in the U.K., in *Farming and the Countryside: An Economic Analysis of External Costs and Benefits,* Hanley, N., Ed., CAB: Oxford, 1991.
23. Powell, J.R., and Allee, D.J., *The Estimation of Groundwater Protection Benefits and Their Utilization by Local Government Decisionmakers*, U.S. Geological Survey, U.S. Department of the Interior, Washington, DC, 1990.
24. Delavan, W.A., Valuing the benefits of protecting groundwater from nitrate contamination in southeastern Pennsylvania, MSc thesis, Penn. State Univ., 1996.
25. Poe, G.L., Information, risk perceptions, and contingent values: the case of nitrates in groundwater, PhD diss., Univ. Wisconsin, Madison, 1993.
26. Giraldez, C. and Fox, G., An economic analysis of groundwater contamination from agricultural nitrate emissions in Southern Ontario, *Can. J. Ag. Econ.,* 43, 387, 1995.
27. Shultz, S.D. and Lindsay, B.E., The willingness to pay for groundwater protection, *Water Resour. Res.*, 26, 9, 1869, 1990.
28. Bergstrom, J.C. and Dorfman, J.F., Commodity information and willingness-to-pay for groundwater quality protection, *Rev. Agric. Econ.,* 16, 3, 413, 1994.
29. Stenger A. and Willinger, M., Preservation value for groundwater quality in a large aquifer: a contingent-valuation study of the Alsatian aquifer, *J. Env. Manage.* 53, 2, 177, 1998.
30. Poe, G.L., Valuation of groundwater quality using a contingent valuation: damage function approach, *Water Resour. Res.,* 34, 12, 3627, 1998.
31. American Water Works Association (AWWA), *Identification and Treatment of Taste and Odors in Drinking Water*, AWWA Research Foundation, Cooperative Research Report, Mallevialle, J. and Suffit, E.H., Eds., Denver, CO, 1987.
32. Brown, A., Farrow, J.R.C., Rodriguez, R.A., Johnson, B.J., and Bellomo, A.J., Methyl tertiary Butyl Ether (MtBE) contamination of the city of Santa Monica drinking water supply, *Proc. Pet. Hydrocarbons and Org, Chemicals in Groundwater,* Houston, NGWA, 1997.
33. Boyle, K.J., Poe, G.L., and Bergstrom, J.C., What do we know about groundwater values? Preliminary implications from a meta analysis of contingent-valuation studies, *Amer. J. Agr. Econ.*, 76, 1055, 1994.
34. Powell, J.R, Allee, D.J., and McClintock, C., Groundwater protection benefits and local community planning: impact of contingent valuation information, *Amer. J. Agr. Econ.* 76, 1068, 1994.

Part II

Applying Economics to Groundwater

5 Overview of Economic Analysis

This chapter provides an overview of the project implementation framework and how economic analysis or appraisal fits within it. Economic appraisal uses information and data input from other decision-making and analysis tools and also could be replaced or complemented by alternative appraisal approaches.

5.1 PROJECT IMPLEMENTATION FRAMEWORK AND ECONOMIC ANALYSIS

Economic analysis involves the definition, measurement, and comparison of costs and benefits of an action where individuals' preferences are used as the rod of measurement and money is used as the unit. Such analysis can be implemented at the strategic level for policy design and assessment and at the tactical level for design and assessment of implementation-related actions. The former, strategic use is possibly the more important use of economic analysis, which should aim to answer two questions in order to produce useful input for decision making:

- Should any action be taken to achieve a given objective? In other words, is the objective worth achieving?
- If so, which action should be taken?

If the first question is answered positively (i.e., the objectives are set), this implies that the benefits of achieving that objective are worth the costs of the actions necessary. This naturally requires direct comparison of costs and benefits or, in economic terminology, *cost–benefit analysis*. The second question of what to do can also be answered by economic appraisal but does not require estimating the benefits of every option of the action. Simply choosing the least costly option that will achieve the objective will suffice. This process is termed *cost-effectiveness* or *least-cost analysis* in the literature.

In the context of groundwater remediation, the distinction between these two levels of economic analysis applies very clearly:

- Setting remediation objective(s) and strategic approaches should involve cost–benefit analysis.
- Selecting a particular remediation technology that meets the objective(s) requires cost-effectiveness analysis.

Further details on the decision-making process in the specific context of groundwater remediation are presented in Part III. This chapter gives a generic overview of these two economic appraisal methodologies. Further details can be found in USEPA[1] guidance on economic appraisal, the HM. Treasury,[2] and many textbooks.[3-7]

Figure 5.1 is a diagrammatic illustration of the steps in what might be described as an idealized decision-making, appraisal, and implementation process.[8] It shows each step of appraisal involved in a more or less chronological order. Different approaches to appraisal differ in their undertaking of all but steps 5 and 6. Section 5.1 presents details on cost–benefit analysis; Section 5.2 on cost effectiveness and Section 5.3 on other types of appraisal. Section 5.4 completes the chapter with an overview of sources of error that can affect all appraisal approaches.

- Step 1 — Define the objectives. An objective is something that decision makers seek to accomplish or to obtain by means of their decision. In the context of this book, remedial or prevention objectives could be to address the source of contamination, the process of contamination, or the receptors potentially at risk of contamination. Chapter 11 discusses establishing and appraisal remediation objectives in more detail.
- Step 2 — Define the baseline and scope. This involves determining the issues or impacts the appraisal will address. So that resources are allocated to best effect, the potential impacts and alternatives investigated should be those considered most significant. In short, scoping affords an opportunity to establish the appraisal's terms of reference clearly at the outset.
  - Determining the baseline is equally, if not more, important. It describes the state of the world without the proposed policy, project, or program and is the measure against which options with the proposed action can be compared. Having a consistent baseline makes the overall appraisal

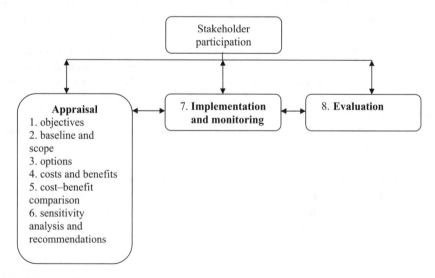

FIGURE 5.1 Economic appraisal and project or policy implementation framework.

consistent. Business-as-usual and do-nothing are the most common baseline scenarios that need to be detailed.

- Step 3 — Identify and screen options in a way that all options could potentially achieve the same objective(s) set in step 1. Otherwise, the outcome of options could not be compared. Options could be at the strategic level, such as remediation approaches, or at the implementation level, such as remediation technologies. Chapter 12 discusses this distinction in further detail. For the purposes of appraisal outline, it is likely that cost–benefit analysis is the more appropriate method for the appraisal of remedial approaches, whereas cost-effectiveness analysis is sufficient for the appraisal of remedial technologies.

- Step 4 — Identify, predict, and assess costs and benefits by considering a range of impact dimensions. These might include, for example, direct, indirect, cumulative, synergistic, permanent, temporary, positive (benefits), and negative (costs) impacts. The significance of the identified impacts can be gauged through expert judgment and dialogue with stakeholders. The main questions to be answered at this stage are: "Who will benefit or lose from different options? And how?"

  - A number of tools or methodologies exist to identify and measure costs and benefits. Two of them are particularly relevant for economic appraisal: cost–benefit analysis (CBA) (see Section 5.2) and cost-effectiveness analysis (CEA) (see Section 5.3). Other appraisal techniques (such as environmental impact analysis), which provide input to CBA and are complementary to both CBA and CEA, are presented in Section 5.4.

- Step 5 — Compare costs and benefits. The appraisal tools differ by the degree to which they "process" impact information and facilitate the comparison of options. For example, qualitative analysis simply presents information on the potential implications of the various options under consideration. Other tools, of which CBA is one, explicitly evaluate the competing options against a common baseline, rank them, and come to a conclusion about the preferred option(s). Appraisal tools therefore differ in the extent to which they explicitly compare and rank options or leave this to decision makers.

- Step 6 — Conduct sensitivity analysis and recommend option(s) at which, in theory, a decision is made as to which of the options under consideration best meets the objectives and should be taken forward. *Sensitivity analysis* refers to the variation in output of the appraisal with respect to changes in the values of the assumptions on which the appraisal is based. Depending on the circumstances, decision makers may have to take a range of other factors into account, in addition to the appraisal findings, in reaching a decision as to the preferred option. If, for any reason, a preferred option cannot be identified, it may be necessary to revisit the options and possibly the objectives.

- Step 7 — Implement the consequences of delivering the preferred option(s) that may be monitored in response to, for example, regulatory requirements or performance targets. Appraisal can support the monitoring process through identifying the most significant potential uses, users, or benefits that might need to be observed into the future and indicate the degree of change (if any) for a given variable.

- Step 8 — Evaluate the performance of the preferred option. The option should be evaluated on the basis of the monitoring findings, and this should be an ongoing process throughout the life of a policy, plan, program, or project. Changes in the overall objectives might be made in light of the review and evaluation.

5.1.1 STAKEHOLDER PARTICIPATION

Key questions to consider in undertaking an economic appraisal are: "Who should be involved? When? And in what capacity?" Ideally, stakeholder involvement should be an integral part of the appraisal process and should help to ensure that the appraisal is transparent and commands a degree of ownership and support. The extent to which stakeholders are involved in appraisal and the way in which they are engaged will vary according to the characteristics of the process and should reflect the significance of the decision at hand.

5.2 COST–BENEFIT ANALYSIS

Well functioning markets are assumed to lead to an efficient allocation of resources. However, when markets fail, as in the case of the absence of markets or the inability of prices to reflect the full economic value of environmental resources, there is a *prima facie* rationale for government intervention. But governments should justify intervention (e.g., setting remediation standards) or demonstrate superior efficiency of their intervention relative to the alternatives. Cost–benefit analysis helps social decision making to facilitate more efficient allocation of society's resources. CBA is also used for private-sector decisions, though to a more limited extent, as shown later.

CBA is an analytical framework used to evaluate public- and private-sector expenditure decisions. It applies to policies, programs, projects, regulations, demonstrations, and other government interventions. It requires systematic counting of all costs and benefits, whether readily quantifiable or difficult to measure. It quantifies as many of the expenditure or actions as possible in monetary terms. Whether in monetary or other terms, it should account for costs and benefits to all members of society.

Box 5.1 illustrates the guidance documents from United States and Europe that recommend the use of economic analysis in general and CBA in particular.

Box 5.1 Inclusion of Environmental Impacts in CBA (Examples from Legislation and Guidance)

Chapter 2 of the U.S. Environment Protection Agency guidance on economic appraisal states that "Policy makers need information on the benefits, costs, and other effects of alternative options for addressing a particular environmental problem in order to make sound policy decisions. In addition, various statutes specifically require economic analyses of policy actions."[1] One such statute is the Executive Order 12866, "Regulatory Planning and Review," which requires analysis of benefits and costs for all significant regulatory actions. This suggests that benefits should justify costs. Benefits include "economic, environmental, public health and safety, other advantages, distributive impacts, and equity" and may not all be quantified. Some commentators suggest that EO 12866 endorses CBA as an "accounting framework" rather than an "optimizing tool."

Article 174 of the Treaty of Europe states that in preparing its policy on the environment, the Community shall take account of:

- Available scientific and technical data
- Environmental conditions in the various regions of the Community
- Potential benefits and costs of action or lack of action
- Economic and social development of the Community as a whole and the balanced development of its regions

The guidance on project appraisal issued by the Treasury in the United Kingdom states that "Appraisal should provide an assessment of whether a proposal is worthwhile, and clearly communicate conclusions and recommendations. The essential technique is option appraisal, whereby government intervention is validated, objectives are set and options are created and reviewed, by analyzing their costs and benefits. Within this framework, cost benefit analysis is recommended, as contrasted wit cost-effectiveness analysis, below, with supplementary techniques to be used for weighing up those costs and benefits that remain unvalued." The guidance defines cost–benefit analysis as "analysis which quantifies in monetary terms as many of the costs and benefits of a proposal as feasible, including items for which the market does not provide a satisfactory measure of economic value." The same guidance defines cost-effectiveness analysis as "analysis that compares the costs of alternative ways of producing the same or similar outputs."[2]

The foreword to the project appraisal guidance prepared for the appraisal staff at the European Investment Bank[9] states that:

All projects funded by the European Investment Bank need to be justified in economic, financial, technical and environmental terms. The Bank's Projects Directorate uses appraisal methods that reflect these various concerns. At the same time, we are aware of the rapid strides being taken in the development of environmental appraisal, and in particular the economic valuation of environmental impacts. The general principles and methods of environmental valuation are now widely known and a large body of empirical evidence is being assembled. The current report is intended to make this academic and research literature more accessible for operational appraisal purposes.... The whole program aims to give project staff the analytical tools and data necessary

to reflect as fully and accurately as possible the environmental impact
of projects funded by the Bank.

The rest of this section provides a brief overview of the main steps of undertaking
a cost–benefit analysis. More detailed and theoretical application of these steps in
the specific context of groundwater contamination and remediation is presented in
Chapter 7, with costs further discussed in Chapter 8 and benefits in Chapter 9, with
a fully developed framework in Part III.

5.2.1 DEFINITION OF COSTS AND BENEFITS

As Chapter 4 showed, the economic value of an environmental resource is determined by individuals' preferences for or against the given changes to that resource.
Following this concept, a *cost* can be defined as a change that people prefer to avoid,
are willing to pay to avoid, or are willing to accept compensation to tolerate.
Similarly, a *benefit* can be defined as a change that people prefer, are willing to pay
to secure, or are willing to accept compensation to forgo. In the context of groundwater remediation, benefits are the environmental damage avoided plus other benefits
of clean-up. Costs, on the other hand, consist of financial and environmental costs
of undertaking remediation.

In order to define and measure costs and benefits in a particular project, first, a
baseline or "without" the project situation must be defined. The most frequently
used baseline or "without" case is the do-nothing scenario. In the context of groundwater contamination, a baseline of do-nothing would assume that no remediation
action takes place.

In some contexts, the do-nothing case may not be a credible option. The other
most popular choice for a baseline is the business-as-usual case, meaning whatever
practice is currently applied or whatever trends are observed continue as before.
Thus, the choice of the baseline depends very much on the objectives of the scheme.

Once the baseline case is identified, costs and benefits of a given project
alternative (the "with" case) can be measured as negative and positive changes
from the baseline (the "without" case), respectively. This definition and measurement of the costs and benefits of a project alternative against a baseline case is
known as the *with–without principle* of CBA.

As discussed in Chapter 4, some components of the total economic value of a
resource (and hence costs and benefits of the changes in the quality and quantity of
that resource) are already reflected in actual markets. If this is the case, market price
(which is itself a lower-bound indicator of WTP) can be used. In the absence of
market prices, willingness to pay and willingness to accept compensation estimates
for those nonmarket costs and benefits are used.

In most applications of CBA, costs and benefits are measured using average
estimates of WTP and WTA, and their comparison on the average is what matters.
However, in some cases, who benefits and who bears the costs matter. In the case
of groundwater remediation, those who use the contaminated water would benefit

from clean-up in terms of avoided cost of alternative water sources, protection of habitats and species, avoided impacts on human health, and so on, and others would bear the financial costs of clean-up.

The fact that different groups benefit or incur costs due to groundwater remediation may matter more if there are socio-economic differences between the two groups. Although in some cases these differences are seen as a political issue or a matter for overall economic policy (such as income distribution) and dealt with separately, in others (especially when spending of public funds is involved), distribution of costs and benefits becomes a concern for the economic analysis.

At the very least, analysis can show how the costs and benefits of each option are spread across different income groups. For example, a proposal providing greater benefits to lower income groups can be rated more favorably than one whose benefits accrue to higher income groups. Her Majesty's (HM) Treasury[2] advises that although in principle distributional impacts should be taken into account, in practice sufficient information is most unlikely to be available at acceptable cost and effort for many applications. The same guidance recommends that the decision about whether an explicit adjustment is warranted should be informed by:

- Scale of the impact associated with a particular project or proposal
- Likely robustness of any calculation of distributional impacts
- Type of project being assessed

In the context of groundwater remediation, distributional impacts of costs and benefits could become a crucial factor in decision making if there is a large discrepancy between the income levels of problem holders and costs of remediation, and the income levels of those affected by contamination and the benefits of remediation that will accrue to them.

5.2.2   COSTS AND BENEFITS OVER TIME

Costs and benefits of any project will be spread over different time periods. This is particularly important for environmental problems like groundwater contamination, which may not be detected for a long time and, once detected, may take a long time to remediate. Cost–benefit analysis of remediation options requires aggregation of all these various costs and benefits for comparison. However, it is generally not possible simply to add benefits and costs as they accrue over time; assessments must account for the timing of costs and benefits.

Discounting is a procedure that weighs present and future benefits and costs so that they may be compared. In general, present gains or losses are weighed more heavily than future gains or losses for two reasons. The first is the productivity of capital argument: one unit of currency that is invested today will be worth more next year due to the interest on capital. Technically, if r is the interest rate, then $1 now and $1 + r$ next year are to be viewed as equivalent. Put another way, $1 next year is the same as $1/(1 + r)$ now. However, some authors have questioned whether this argument can be applied to natural capital, that is, environmental resources and services provided by them.

The second reason, the so-called time preference argument, relies on the simple observation of the behavior of individuals and concludes that, regardless of interest rates, people do prefer their benefits now rather than later, simply because people have limited lifetimes and hence are impatient. If what matters are individuals' preferences, their preference about the incidence of costs and benefits through time cannot logically be excluded. In turn, this means that future benefits and costs must be discounted. The general formula for discounting benefits is given in Equation 5.1 and Equation 5.2:

$$PV(B_t) = \frac{B_0}{(1+r)^0} + \frac{B_1}{(1+r)^1} + \frac{B_2}{(1+r)^2} + \ldots + \frac{B_n}{(1+r)^n} \qquad (5.1)$$

or

$$PV(B_t) = \sum_{t=0}^{t=n} \frac{B_t}{(1+r)^t} \qquad (5.2)$$

where $PV(B_t)$ is the present value of all future benefits, B_t is the benefit that occurs in time period t, n is the project lifetime, and r is the discount rate.

The formula for discounting costs is similar, as shown in Equation 5.3:

$$PV(C_t) = \sum_{t=0}^{t=n} \frac{C_t}{(1+r)^t} \qquad (5.3)$$

where $PV(C_t)$ is the present value of all future costs, C_t is the cost that occurs in time period t, n is the project lifetime, and r is the discount rate.

The choice of discount rate is open to some debate among economists and is not straightforward. A concern is that discounting effectively devalues the future by putting an inordinate emphasis on present value. This is especially the case for natural resources with long gestation periods.[10] Deep aquifers, which contain relatively old water and are recharged slowly, are an example. A high discount rate could result in an economic analysis that promotes unsustainable abstraction rates, based on the relatively high value of water in the near term. In the same way, a high discount rate may mean that the benefits of protecting groundwater for the future are too small to warrant expenditure in the near term, resulting in a deferral of action. Indeed, high discount rates are frowned upon in much of the environmental literature because they tend to shift the burden of responsibility for environmental protection and remediation onto future generations.[10]

Therefore, it could be argued that because we are measuring the costs and benefits as they accrue to the society as a whole, the time preference of society is what counts, rather than the time preference of an individual. Because the lifetime of a society is much longer than that of a single individual, the discount rate used for the society (i.e., the *social discount rate*) should be lower. However, estimating the rate of time preference for the society as a whole is not straightforward. Pearce and Ulph showed that the rate at which future costs and benefits should be discounted is about 2.5 to 3%.[11] Freeman derived a similar estimate of 2 to 3% for the United States.[12] Several authors have found that individuals have lower discount rates for environmental (or nonmarket) goods than for market goods.

Guidance on appropriate social discount rates differs among government institutions and is generally higher. For example, HM Treasury currently advises that the rate should be 3.5%.[2] However, the guidance also addresses cases where the appraisal depends materially on the discounting effects in the very long term and recommends that lower discount rates should be used for the longer term (beyond 30 years). The main rationale given in the guidance for declining long-term discount rates is to take uncertainty about the future into account. In light of the evidence, the guidance recommends to use 3.5% for the first 30 years, 3% for years 31 through 75, 2.5% for years 76 through 125, 2% for years 126 through 200, 1.5% for years 201 through 300, and 1% for years 301 and beyond.

USEPA recommends that, for projects and policies with intragenerational impacts, a discount rate of around 2 to 3% is justified.[1] The EPA guidance states that, for projects and policies that have long-term (intergenerational) impacts, economic analyses should generally include a no-discounting scenario by displaying the streams of costs and benefits over time. This is not equivalent to calculating a present value using a discount rate of zero (i.e., the flow of benefits and costs should be displayed rather than a summation of values). Both the U.S. and the U.K. guidance documents recommend sensitivity analysis, which shows the effect of different discount rates on the present value of costs and benefits and, hence, on the ultimate decision.[1,2]

5.2.3 COMPARING COSTS AND BENEFITS

Calculating the net present value (NPV) is one method of comparing costs and benefits. The relevant formula is shown in Equation 5.4:

$$NPV = PV(B_t) \square PV(C_t) = \sum_{t=0}^{t=n} \frac{B_t \square C_t}{(1+r)^t} \qquad (5.4)$$

NPV provides a simple formula for assessing whether a decision, action, project, or policy is worthwhile. If NPV is positive, benefits exceed costs and overall the project is worthwhile. If there are a number of project alternatives with positive NPVs, the one with the highest NPV should be implemented.

Equation 5.4 does not include any reference to distributional impacts of costs and benefits. If they are to be included, the equation should be revised as in Equation 5.5:

$$NPV = \sum_{i=1}^{i=p} \sum_{t=0}^{t=n} \frac{a_i(B_{it} \square C_{it})}{(1+r)^t} \qquad (5.5)$$

where a_i is the weight to be attached to the ith income group and is calculated as $(\overline{Y}/Y_i)$, or the ratio of the average income (of the nation or the affected community) and the income of the ith income group, and p is the total affected population. The distributional weight becomes less than 1 for income groups above the average and hence, when multiplied with benefits and costs accruing to that group, reduces their relative value. On the other hand, the weight becomes more than 1 for income groups below the average, inflating the relative value of costs and benefits that accrue to that group.*

The benefit–cost ratio is another way of comparing the discounted (present) value of costs and benefits, as shown in Equation 5.6:

$$Benefit \, / \, Cost = \frac{PV(B_t)}{PV(C_t)} \qquad (5.6)$$

A benefit–cost ratio greater than 1 means that, over time, benefits are greater than costs, so projects with benefit–cost ratios greater than 1 are worthwhile, according to CBA. With a fixed budget, the benefit–cost ratio is the right indicator for setting priorities among options. Maximum benefits are, therefore, obtained by first implementing the option with the highest benefit–cost ratio, then the second highest, and so on. Once the budget limit is reached, the remaining projects cannot be implemented even if they have benefit–cost ratios greater than 1.

The process of selecting the best alternative among several for implementing a project is therefore relatively straightforward, as long as all costs and benefits are known and can be expressed in the same unit (i.e., money in this context). However, as discussed in Chapter 4, it is likely that there will be benefits and costs from projects that cannot be expressed in monetary terms. This is especially prevalent in cases like groundwater contamination, where not only cost and benefit information is missing or difficult to obtain but so is scientific and technical information.

This makes it inevitable that CBA will define, measure, and compare some costs and benefits in monetary terms, whereas others can only be expressed in their original physical units or even in qualitative terms. In this case, the procedure should be to compare all monetized costs and benefits as previously and to list the nonmonetized effects.** Note that a nonmonetized negative indicator constitutes a cost and a nonmonetized positive indicator constitutes a benefit. There are four ways of dealing with mixed outcomes:

---

* This weight assumes that a marginal (extra) unit of income is valued the same by different income groups or that the income elasticity is unity. This is unlikely to be the case, because poorer income groups are likely to value extra income more highly than the richer groups. If so, the weight formula becomes $(\overline{Y}/Y_i)^b$, where b is the income elasticity of demand and is usually less than one. Although there is little literature about this, a value of 0.35 for b is sometimes used.

** USEPA guidance suggest that in some cases, nonmonetary costs and benefits should also be discounted. This is especially recommended when it is thought that not discounting these costs and benefits would lead to postponement of investments, because costs can be monetized but benefits cannot. In the U.K., this is not the practice, and it is not covered further in this book. Interested readers are referred to Chapter 6 of the EPA Guidance.[1]

- If monetized benefits (B_m) exceed monetized costs (C_m) and the nonmonetized indicators are judged mainly to be positive, then proceed, because the benefits more than outweigh the costs.
- If monetized benefits exceed monetized costs and the nonmonetized indicators are judged mainly to be negative, then compare net monetized benefits with the nonmonetized costs (C_{nm}). Using professional judgment, ask whether the nonmonetized costs are likely to be greater than the net monetized benefits. If they are, the option is not worthwhile. If they are not, the option is potentially worth pursuing.
- If monetized costs exceed monetized benefits and the nonmonetized indicators are judged mainly to be positive, then compare net monetized costs with the nonmonetized benefits (B_{nm}). Using professional judgment, ask whether the nonmonetized benefits are likely to be greater than the net monetized costs. If they are, the option is potentially worth pursuing. If they are not, the option is not worthwhile.
- If monetized costs exceed monetized benefits and the nonmonetized indicators are judged mainly to be negative, then the scheme is not worth pursuing.

Table 5.1 summarizes these four possible outcomes. The incompleteness of economic analysis information should be accepted as it is, and gaps should be filled with further research whenever possible. The view that if we cannot monetize

---

TABLE 5.1
The Four Potential Outcomes for Mixed Monetary and Nonmonetary Assessment

	Bm > Cm	Bm < C m
$B_{nm} > 0$	1. Proceed because benefits more than outweigh costs.	3. Judge whether $B_{nm} > [C_m - B_m]$. If so, proceed. Judge whether $B_{nm} < [C_m - B_m]$. If so, reject.
$C_{nm} > 0$	2. Judge whether $[B_m - C_m] > C_{nm}$. If so, proceed. Judge whether $[B_m - C_m] < C_{nm}$. If so, reject.	4. Reject because costs more than outweigh benefits

Note: *m* denotes monetary estimates and *nm* denotes nonmonetary indicators. All monetary measures should be considered in NPV terms.

everything, then nothing should be monetized should be avoided. This view amounts to rejecting valuable information about people's preferences.

Section 5.4 provides an overview of alternative approaches to decision-making analysis or project appraisal. Some of these can only generate information for CBA because they are designed to collect information, rather than process it for decision-making purposes. Alternative approaches process cost and benefit information using criteria other than individuals' preferences. As Section 5.4 shows, these approaches also have their shortcomings.

5.2.4 RISK, UNCERTAINTY, AND SENSITIVITY ANALYSIS

The assumptions underlying the economic analysis are likely to have an element of risk or uncertainty attached to them. *Risk* is defined as some known combination of the probability that an event will occur and the scale of the event. *Uncertainty* arises when the probability distribution is not known and the scale of the event, if it occurs, may be known accurately or only imperfectly. The distinction between risk and uncertainty can be important, because the means of dealing with each may well be different. Both risk and uncertainty here refer to financial/economic as well as technological and environmental aspects of groundwater remediation (see also Chapter 3 and Chapter 6). Risk and uncertainty in the specific context of groundwater are discussed in Chapter 3 and Chapter 6.

5.2.4.1 Risk

As far as financial risk is concerned, market interest rates often incorporate this: where repayment prospects are more dubious, the interest rate will be correspondingly greater. However, it is preferable in project appraisal to examine risk more systematically, for example, through sensitivity analysis. This is particularly true if the risks of different projects are not highly correlated and are "pooled" across society, rather than impacting primarily on a single group. It is worth bearing in mind, in this context, that risks for a problem holder need not represent equivalent risks, or even any risk, for society as a whole (one individual's loss can be another's gain). This fact has important implications for the debate between public and private provision of groundwater remediation and the level of regulation.

If an event with a cost valued at $100 occurs with a probability of 10% (e.g, there is a 10% chance that groundwater contamination will reach a borehole and cause $100 worth of damage), one approach might be to multiply the two numbers so that risk equals $10 This is an example of an *expected value* approach to representing risk. Equation 5.7 shows how the net present value estimate would change when expected, rather than absolute, values for costs and benefits are used.

$$E(NPV) = PV(B_t) \square PV(C_t) = \bullet \sum_{t=0}^{t=n} \frac{(p_b \square B_t) \square (p_c \square C_t)}{(1+r)^t} \qquad (5.7)$$

where $E(NPV)$ is the expected value of the net present value, p_b is the probability of benefits occurring, and p_c is the probability of costs occurring.

The linear interpolation involved in Equation 5.7 is probably inappropriate at the individual level, because individuals may be risk averse in most settings. Risk aversion would suggest that the value of the 10% risk of losing $100 is greater than $10 — an individual might be prepared to pay $12, $15, or more to avoid the risk of the greater loss. There is nothing irrational in this; it is a perfectly normal expression of preferences. Indeed, it is the whole principle underlying the insurance market — if people were not willing to pay more than the actuarial value for insurance, no insurer could ever make a profit. But risk aversion is context specific: in some settings, individuals are risk loving (and to continue the analogy, if they were not, no bookmaker could ever make a profit).

The effects of risk aversion may be mitigated at a social level, to the extent that risks are pooled in an independent way. That is, if the risks of several different projects (e.g., risk of failure of remediation, risk of contamination spreading) are statistically independent, then the overall level of risk in the portfolio (viewed as the standard deviation around the expected return) may be much lower, relative to the return, than for each individual project. In such cases, the use of the expected value is less problematic. However, where the risks are not independent but are linked, this will not hold, and societal preferences may very well be risk averse. This is likely for risks that may have a broad impact right across the economy, global warming being one example.

5.2.4.2 Uncertainty

Most environmental decision problems are likely to be characterized by uncertainty rather than risk. In the case of uncertainty, it may be known that there is a possibility of a $100 loss, but the probability of that loss is not known (e.g., the economic value of a borehole is $100 which may be lost if water is contaminated but the risk of contamination is not known). Or it may be that the scale of the event is known in only qualitative terms, and the probability is not known at all.

The simplest approach to dealing with uncertainty is to adopt sensitivity analysis. This involves showing how the outcome of the CBA varies according to the adoption of different values for some key parameters. In other words, it allows what-if scenarios to be tested. Sensitivity analysis consists of repeating the analysis by changing the value of a single key parameter at a time and comparing the resulting NPV or benefit–cost ratio with the original analysis. In addition to testing the effect of changing individual parameters, combinations of assumptions may also be tested.

Sensitivity analysis by itself resolves nothing: it simply shows the sensitivity of the cost–benefit calculation to changes in assumed values of parameters. However, this has the advantage of focusing attention on the values of the parameters in question, including:

- Minimum possible values
- Best estimates, which are usually used in the main CBA rather than in the sensitivity analysis
- Maximum possible values

Several situations might emerge:

- Benefits exceed costs for the project, regardless of the value chosen for the key parameter. Then the result is robust.
- Costs may exceed benefits for the project regardless of the value chosen for the key parameter. Again, the result is robust.
- Project may pass (or fail) a cost–benefit test for some values of the parameter but not for others. This outcome forces the decision maker to express a judgment about which value of the parameter is most likely. Effectively, an uncertainty problem is converted to something akin to a risk problem by the assignment of judgmental probabilities.

One approach to incorporating uncertainty into decision making is to use a payoff matrix, which, in its simplest form presented here, does not involve any sophisticated modeling exercise but a framework for making professional judgments about uncertain situations. If the objective of the analysis is to maximize net benefits, as is the case for a CBA, the numbers in the payoff matrix record values of net benefits. These net benefits depend both on what decision (D) is taken (e.g., D_1 involves a groundwater remediation investment and D_2 involves no remediation) and on what the *state of the world* (S) is (e.g., different levels of water availability in the future). The state of the world simply reflects the possibilities that may occur in the future. The pay-off matrix in Table 5.2 shows the net present values of decisions D_1 and D_2 in the states of the world S_1 and S_2. These can be estimated using sensitivity analysis, (i.e., rerunning the CBA with different assumptions). However, the probabilities attached to the states of the world and hence the outcomes of decisions D_1 and D_2 are not known.

In Table 5.2, if S_1 occurs, the best decision is D_1. But choosing D_1 is risky because S_2 could occur and there could be a loss of $45 (see below). The following decision rules are possible depending on whether the decision maker is risk loving or risk averse, respectively:

- Maximax — Choose the option that maximizes benefits (here, D_1 with +$100). This criterion would be chosen by an optimist because there is a risk that S_2 would occur and losses would be incurred.
- Maximin — Choose the option that minimizes losses (here D_2 with +$30). The minimum payoffs are −15 and +30, so the decision maker maximizes these minima. The decision maker using this criterion is cautious: he or she avoids the worst outcomes.

TABLE 5.2
Pay-off Matrix

	State of the World (1)	State of the World (2)
Decision 1	+$100	−$15
Decision 2	+$90	+$30

Other criteria focus on what would happen if the wrong decision were made. To determine this, a regret matrix should be constructed, as shown in Table 5.3. The regret payoff is defined as the difference between what is actually secured and what could have been secured had the correct decision been made. For example, choosing D_1 with S_1 occurring involves no regret because D_1 has the highest payoff. Choosing D_1 with S_2 occurring involves forgoing $30 (had the correct decision, D_2, been made) and losing $15, a regret of $45. Choosing D_2 in S_1 yields $90, but had D_1 been chosen it could have been $100, so the regret is $10. Choosing D_2 in S_2 involves getting $30, but choosing D_1 in S_2 would have produced −$15, so the regret is zero.

A criterion for choice is now *minimax regret*. This involves taking the maximum regrets from the regret matrix ($10 and $45) and minimizing these (choosing $10), that is, D_2.

5.3 APPLYING COST–BENEFIT ANALYSIS

So far, this chapter has discussed why and how cost benefit analysis can and should be implemented to inform decision making. This section looks at the two different levels of CBA application in practice (financial vs. economic analysis) and the various obstacles to wider use of economic analysis.

5.3.1 FINANCIAL VS. ECONOMIC ANALYSIS

There are at least two important stakeholders in the context of making remediation decisions. The first is the problem holder, which is usually a private-sector company, or in any case, a single polluter or group of polluters. The other is the regulatory agency that represents the interests of the rest of the society. Whether using economic approaches or not, these two groups undertake their own analyses, concentrating on the costs and benefits that concern themselves alone.

However, the economic analysis of the remediation decision presented in this book looks at groundwater remediation in a more holistic way, in that the costs and benefits are defined from the perspective of the whole society, that is, both the polluters or problem holders and the rest of the society. Therefore, it is interesting to see how economic analysis would be undertaken by the problem holder — the so-called financial or private analysis — and how it should be undertaken as recommended here — the so-called economic or social analysis.

TABLE 5.3
Regret Matrix

	State of the World (1)	State of the World (2)
Decision 1	$0	−$45
Decision 2	−$10	$0

Table 5.4 shows the main points where the two levels of economic analysis differ. The following text summarizes the main points listed in the table.

- Costs — In financial analysis, the only costs of concern are those that are incurred by the problem holder. These include the financial costs of clean-up, for example. In an economic analysis, in addition to these financial costs, costs to the third parties are also included. Examples of these third-party costs (or, to use Chapter 4's terminology, external costs) include blight effect of contamination on surrounding properties, impacts on flora and fauna, and impacts on human health. Further detail on the costs of groundwater contamination and the external costs of remediation can be found in Chapter 8.
- Benefits — In financial analysis, the only benefits of concern are those that accrue to the problem holder. These include the financial returns from the sale of a property that fetches a higher price after clean-up, for example. In an economic analysis, in addition to these financial benefits, benefits to the third parties are also included. Examples of these third-party benefits (external benefits) include avoided blight effect of contamination on surrounding properties and avoided impacts on human health, flora, and fauna. Further detail on the private and external benefits of groundwater of remediation can be found in Chapter 9.
- Market prices vs. shadow prices — Financial analysis includes market prices alone. Because impacts on third parties are excluded from this analysis, nonmarket costs or benefits are also excluded. Market prices include taxes and subsidies as relevant. Because these are paid by and to the problem holders, they are included in the financial analysis. In economic analysis, on the other hand, taxes and subsidies form part of transfer payments. *Transfer payments* refer to changes that benefit some groups (in the case of taxes, the government) and directly cause a loss to other groups (again for taxes, taxpayers) and hence constitute no net change in society's wealth or welfare. Therefore, market prices net of taxes and subsidies, which are known as real or shadow prices, are used in economic analysis. In addition, economic analysis also includes nonmarket costs

TABLE 5.4
Financial and Economic CBA

Financial/Private Analysis	Economic/Social Analysis
Own costs (expenditures)	All costs to society
Own benefits (revenues)	All benefits to society
Market prices (with transfer payments)	Shadow/real prices (market prices wihout transfer payments plus nonmarket costs and benefits)
Private discount rate	Social discount rate
No account of equity impacts	Can account for equity impacts

and benefits because most of the impacts on third parties are, in fact, nonmarket or external to actual markets.

- Discount rate — Both levels of analysis use discounting. However, they differ in the discount rate applied. Financial analysis that is solely concerned with the costs and benefits affecting the problem holder use *private discount rate,* which reflects private cost of capital, risks undertaken, and private time preferences. Economic analysis, on the other hand, uses *social discount rate,* which reflects the ability of the society as a whole to distribute risks associated with individual projects and the lifetime of societies and associated time preference. As a result, private discount rate is higher than social discount rate.
- Equity impacts — In a financial analysis, the question of who benefits or loses (other than the problem holder) is of no consequence, so equity adjustments are not undertaken. In economic analysis, on the other hand, such adjustments may or may not be undertaken, depending on the requirements of the context and the decision maker.

5.3.2 Obstacles to Wider Use of Cost–Benefit Analysis

A fully comprehensive economic analysis, or CBA, requires time, skill, and money. For example, USEPA spent about $1 million on a CBA of regulations reducing lead in gasoline. More typically, major CBA of large EPA projects ($100 million and over) cost around $700,000. But, like anything, these should be judged against the proportionality rule: they are small sums compared to the size of the expenditures they appraise, not to mention the cost of making a wrong decision. The cost of a CBA for a remedial action in a single site with a single-source contamination problem would, of course, cost much less. In fact, CBA may not even be necessary if all parties concerned agree on the remedial objective.

However, the cost of undertaking a CBA is not the only obstacle to its wider use, in general or in the context of groundwater remediation. The philosophy of CBA, which relies on individuals' preferences, gives further clues to why this is so. These obstacles to the application of CBA can be classified as issues of philosophy and content and issues of process.[13]

5.3.2.1 Philosophy and Content

- Credibility — CBA is a quantitative technique, and the resulting quantities are often uncertain. Decision makers are generally averse to uncertainty. However, uncertainty is inevitable in the real world and implications of many decisions can be assessed better with a quantitative technique like CBA than its more qualitative alternatives.
- Morality — CBA uses money values, and there is often a moral hostility to using money as the measuring rod. However, money is chosen as a measuring rod simply because it is a familiar unit for all, it is divisible, and it makes environmental impacts (which are in disparate units) comparable with each other and with financial costs and benefits.

- Efficiency focus — CBA has economic efficiency as its goal. But governments have multiple objectives, hence, CBA appears to be partial and noncomprehensive. CBA is only one of the approaches to decision making, and other factors deemed important can also be taken into account. In addition, equity issues, in terms of identifying who benefits and loses from a particular decision, can be included within CBA.
- Democratic principle — Although it may seem odd to suggest that decision makers oppose democracy, there are concerns about the legitimacy of reflecting preferences in all contexts. More fundamentally, some critics object that, by focusing on the preferences of individuals, economic evaluation takes account only of self-interest. If an individual has a preference for or against something, it might appear that the preference is formed on the basis of what that individual judges to be best for him- or herself. A short way to express this point is to say that the individual acts out of self-interest. Indeed, this is how consumer sovereignty or economic rationality is often characterized. But the issues for which monetary valuation will be used are often those where the public interest is the issue, that is, what is best for society as a whole. This is why the total economic value concept includes (and economic valuation techniques seek to quantify) motives other than self-interest, such as nonuse values.

5.3.2.2 Process

- Flexibility — Some decision makers may feel that by imposing rules, processes, and criteria, CBA compromises their flexibility in decision making. However, in most cases that involve nonmarket costs and benefits, CBA is incomplete due to lack of physical or economic data, so a level of flexibility and decision-making initiative is not only useful but also necessary.
- Participation — Although economic values are based on individuals' preferences, CBA is sometimes criticized for being nonparticipatory, because participation, in the sense of measuring preferences, is not seen as transparent and inclusive enough.
- Capacity — CBA requires a certain level of expertise on the part of those using it or judging its results. Although the framework developed in Part III of this book cannot eliminate the need for economists' expertise, it is aimed to help experts in different fields (including economists) understand each other more clearly.

5.4 COST-EFFECTIVENESS ANALYSIS

Cost-effectiveness analysis (CEA) is different from CBA only in that benefits of remediation are not monetized. In the words of Mishan: "the analysis of cost-effectiveness can be described as a truncated form of cost benefit analysis: it draws inspiration and guidance only from the cost side — or alternatively, only from the benefit side — of a cost benefit format."[4] CEA requires that a level of remediation for a given site is agreed due to regulatory, political, or social reasons. It then becomes a framework to assess the different approaches or technologies to achieve this given level of remediation by comparing the costs of the alternatives.

This least-cost analysis can be transformed into one that searches for the most-benefit analysis. Here, instead of finding the cheapest way to achieve a given objective, we are given a limited budget and asked to spend it in a way that will generate the highest benefit. Formally, both analyses serve the same economic principle: whether the cost or the benefit is the constraint, we are to get the most for our money or, what comes to the same thing, to spend the least for our benefits.

When benefits of remediation are difficult or impossible to estimate in monetary terms, CEA becomes especially useful. However, CEA can answer the question of how to remediate, but it cannot answer the crucial question of whether the level of remediation given is the socially optimal level.

5.5 OTHER APPRAISAL APPROACHES

The main approach advocated in this book is economic tools, including cost–benefit analysis and cost-effectiveness analysis. However, many other appraisal approaches exist, all of which aim to aid the decision-making process. Some of these approaches collect information and present it as input to other approaches. For example, environmental impact assessment generates information that can be used on its own or as an input to CBA. Some approaches both collect and process information. Examples of this last group of approaches are risk–benefit analysis and the multicriteria approach (MCA).

The linkages among various analysis techniques (including CBA, CEA, and MCA), the initial environmental assessment (or site characterization), and risk assessment are shown in Figure 5.2. This illustrates the integral relationship between the technical data collection and risk analysis stages and the economic analysis. The figure does not include other approaches, such as strategic environmental assessment. Thus, it is not exhaustive but focuses on the techniques that are used or that can potentially be used in the context of groundwater protection and remediation. The remainder of this section presents these other appraisal approaches for completeness (for further details, see eftec[14]).

- Environmental impact assessment (EIA) — EIA is an assessment of the impact of a planned activity on the environment. In essence, it is a systematic process whereby information about the environmental effects of an action is collected and evaluated, with the conclusions being used as a tool in decision making. The approach provides information about the physical quantities of environmental impacts, which is used in the CBA. The advantages of the assessment include systematic consideration of environmental consequences of actions and identification of measures needed to mitigate serious negative impacts. However, EIA does not

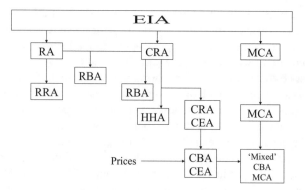

FIGURE 5.2 Appraisal approaches and how they link to economic analysis.

provide a decision rule or procedure for aggregating disparate environmental impacts.

- Risk assessment (RA) — RA involves the estimation of the probability and severity of hazards to human health, safety, and ecosystem functioning or health. Any hazardous substance has the potential to cause these forms of harm. The RA assesses the potential hazard, the likelihood of that hazard's being realized, and the severity of the impact for any given level of exposure (see Chapter 3). Here, however, RA is viewed as a decision-making or appraisal tool including economic and financial solutions. Risk assessment can take a number of different forms, depending on the type of risk assessed and the way risks (and indeed benefits) are compared. In this book, the RA is a critical input in the economic appraisal.

- Comparative risk assessment (CRA) — CRA involves allocating risk-preventing resources among risks in an efficient manner. Cost–risk ratios enable an allocation of resources that ensures that the maximum risk reduction is made per dollar spent. CRA may be applied satisfactorily to one risk (e.g., health risk), but when more than one risk is involved (e.g., health and ecosystem risks), problems arise unless the relative weight of each risk is known.

- Risk–benefit analysis (RBA) — RBA can be interpreted as CBA (or risk-cost-benefit) when policies have monetary costs, risks, and benefits. This creates the decision-making rule of (Benefits − costs − risks) < 0. RBA can also be used to compare risks to benefits in a standardized form other than monetary. An example is measuring benefits of traveling in time per saved passenger kilometers, which allows a measure of risk per passenger kilometer, giving a risk-to-benefit ratio where a policy with a lower ratio is chosen.

- Risk–risk analysis (RRA) — RRA evaluates policy in terms of the reduced risks rather than the risks that would have occurred without the policy or project. This allows the inclusion of effects that may increase the risks of an action that may not have been present without the policy or project. For example, a policy enforcing seatbelts will prevent a given amount of risk; however, the policy may encourage faster driving and, therefore, increase risk. RRA would use the latter measurement of risk reduction rather than forgone risks, allowing the incorporation of behavioral response.

- Health–health analysis (HHA) — HHA incorporates an income–mortality effect into measuring the benefits of regulation. The assumption that the risk of mortality is greater at low income than at high income implies that the reduction in income due to the cost of regulation may increase mortality. If the cost of regulation that reduces mortality is $10 billion, an equivalent total reduction in household income is implied, and this increases the risk of mortality. If the increase in mortality from the reduction in income is greater than the decrease in mortality from the regulation, the regulation fails HHA and should not be undertaken. It is possible for a regulation to pass HHA but fail CBA, so HHA should not be the only analysis undertaken.

- Multicriteria analysis (MCA) — MCA is a two-stage procedure. The first stage identifies a set of goals or objectives (e.g., different levels of remediation) and seeks to identify the trade-offs between those objectives as well as trade-offs between different ways of achieving a given objective (e.g., different engineering alternatives for remediation). The second stage seeks to identify the best policy by attaching weights to the various objectives. Although MCA does not require the monetization of objectives, it does require a rational analysis of what must be traded off for what. Thus, a set of weights used in MCA can be the monetary values of financial, environmental, and social costs in question. Therefore, MCA is capable of combining monetary and nonmonetary values for different costs and benefits.

The important feature of MCA is that it embraces objectives that CBA appears not to embrace (or not to embrace as a matter of course). For example, it could include a distributional objective (fairness, equity); some assessment of sustainability; and wider national concerns, such as competitiveness, employment, regional balance, and so on.

If MCA is wider than CBA, why not recommend MCA over CBA? The question is somewhat misleading, because CBA is in fact a particular form of MCA.[13] There is nothing in MCA that says efficiency is not important and nothing that says impacts should not be monetized where appropriate. If there is an equity goal, this may not be suited to monetization, in which case something that is more efficient but less equitable must be traded against something that is less efficient and more equitable. The efficiency status cannot be determined, however, without some form of CBA. Thus CBA can, and should, be an input into MCA.

Proceeding to MCA without CBA can entail a number of problems, including:

- Many MCAs do not account for public preferences at all but use expert judgment. This runs counter to the democratic principle already introduced.
- MCAs face considerable difficulties with time discounting and changes in relative values.
- MCAs often risk double-counting of objectives.

REFERENCES

1. USEPA (U.S. Environmental Protection Agency), *Guidelines for Preparing Economic Analyses*, Office of the Administrator, USEPA, Washington, DC, 2000.
2. HM Treasury, *The Green Book: Appraisal and Evaluation in Central Government*, TSO, London, 2003.
3. Pearce, D.W., *Economic Values and the Natural World*, Earthscan, London, 1993.
4. Mishan, E.J., *Cost-Benefit Analysis: An Informal Introduction*, 4th ed., Routledge, London, 1994.
5. Layard, R. and Glaister, S., *Cost–Benefit Analysis*, 2nd ed., Cambridge University Press, Cambridge, U.K., 1994.
6. Dixon, J.A., Scura, L.F., Carpenter, R.A., and Sherman, P.B., *Economic Analysis of Environmental Impacts*, Earthscan, London, 1994.
7. Hanley, N. and Spash, C., *Cost–Benefit Analysis and the Environment*, Edward Elgar, England, Cheltenham, 1995.
8. Environment Agency, *Integrated Appraisal Methods*, R&D Technical Report E2-044/TR, Bristol, 2003.
9. IVM and eftec et al., *Framework to Assess Environmental Costs and Benefits for a Range of Total Water Management Options*, R&D Technical Report W156, Environment Agency, London, 1998.
10. Pearce, D.W. and Warford, J.J., *World without End*, World Bank, Oxford University Press, Oxford, 1993.
11. Pearce, D.W, and Ulph, D., A social discount rate for the United Kingdom, in *Economics and Environment: Essays on Ecological Economics and Sustainable Development*, Pearce, D.W., Ed., Edward Elgar, Cheltenham, U.K., 268, 1999.
12. Freeman, M.A., *The Measurement of Environmental and Resource Values: Theory and Methods*, Resources for the Future, Washington, DC, 1993.
13. eftec and RIVM, *Valuing the Benefits of Environmental Policy: The Netherlands*, RIVM report 481505 024, Bilthoven, 2001.
14. eftec, *Review of Technical Guidance on Environmental Appraisal*, DETR, London, 1998.

6 Groundwater in an Economic Context

6.1 OVERVIEW

Assigning costs and benefits to groundwater remediation will depend in part on our technical understanding of the problem. An understanding of the nature, type, and distribution of the contamination and the fundamentals of the groundwater regime is required to calculate risks and to assess and quantify damage. Knowledge of future movement of the plume can be used to calculate future expected impacts and thus predict future damage. Estimating the benefits of remediation depends in part on a view of which ecosystems and surface water bodies will be spared if remediation occurs. This chapter provides a discussion of the main relevant technical hydrogeological issues that bear on the use of economic analysis as a tool in determining the degree to which groundwater should be remediated and in choosing the best approach and method to achieve the remedial objectives.

6.2 REMEDIAL OBJECTIVES, APPROACHES, AND TECHNOLOGIES

In developing a methodology for applying economic analysis to groundwater problems, clear terminology is vital. There exists in the literature today no single set of terms that clearly defines the various stages of remedial design and decision making. Clear distinctions among the different levels at which remedial decisions are made are required if costs and benefits are to be assigned to competing options as part of an economic analysis. For this reason, the following terms are defined and adhered to throughout this book:

- *Remedial objective* describes the overall intent of the remediation project. Objectives could include the degree to which groundwater is to be remediated, the protection of specific receptors, or the elimination or reduction of certain unacceptable risks. Remedial objectives are limited in number and are based on receptor protection. Examples of remedial objectives include protecting a receptor from future damage, preventing additional damage to an existing receptor, and making a site fit for some future purpose, such as redevelopment.
- *Remedial approach* is the conceptual manner in which the objective is to be reached. In this book, remedial approaches refer specifically to measures that break the source–pathway–receptor (SPR) linkage, either by removing part or all of the source of contamination, by cutting the path-

way, or by isolating or removing the receptor. Examples of remedial approaches include physical containment, source removal, receptor isolation, natural attenuation, and monitoring.

- *Remedial technologies* are the specific tools that form the components of the approach. For example, physical containment can be achieved through use of slurry walls, sheet pile walls, or liners, often in conjunction with groundwater pumping and treatment. Source removal can be achieved through excavation and on-site treatment of contaminated soils (by a variety of techniques) or through many available *in situ* techniques. A remedial solution often involves the use of several different remedial technologies.

Remedial objectives should be known before detailed design (technology selection) occurs. The choice of a remedial approach is the critical intermediate step, which can be used as a tool to help set objectives (by considering and comparing various approaches at the conceptual level) and to guide the selection of the technological components that will make up the final design. As will be seen in the chapters that follow, the remedial approach is the level at which comparative economic analysis can most readily be carried out.

6.3 GROUNDWATER, RISK, AND UNCERTAINTY

6.3.1 RISK

As discussed in Chapter 3, the environmental risks associated with groundwater contamination can be classified into three categories:

- Risks of damage to groundwater resources themselves (aquifers), and thus to the users of that groundwater (humans, crops, animals)
- Risks of impact to surface water resources, as a result of groundwater's contribution to the resource (via baseflow discharge), and thus to the users of the surface water (humans, crops, animals, ecosystems)
- Risks of impact to receptors as a result of contaminant migration via groundwater (as a risk pathway), including ecosystems, property, natural amenity features, and possibly humans and animals

Because groundwater and the contaminants within it are mobile, impacts may occur at substantial distances from the contamination's original source. Due to the heterogeneity of geological materials, the patterns and velocities of contaminant movement in groundwater are difficult to predict, and there is significant uncertainty involved in any prediction of future impacts.

If multiple sources are involved, as could be the case in an industrial area, commingled plumes could result. Several such situations have been documented in the United States.[1] These could involve several sources, with several different responsible parties. The issues of apportioning responsibility for the damage and assigning costs for remediation to each party are fraught with complications.

Analysis of the risks associated with groundwater contamination will usually involve application of the source–pathway–receptor (SPR) concept, discussed in Chapter 3. For a potential risk to exist, a complete SPR linkage must exist. Estimation of the costs of damage associated with a particular risk will involve some level of analysis of the risk's probability and the likely impacts to the receptor.

6.3.2 Uncertainty and Multiple Risk Linkages

Uncertainty is introduced into the environmental risk analysis through the following:

- Uncertainty in prediction of contaminant behavior within the subsurface (distribution and concentration of contaminants, migration direction and velocity, the effects of retardation and attenuation mechanisms)
- Incomplete site characterization information due to limited resources, leading to uncertainty of information regarding source concentration, mass, and composition
- Assumptions required in formulating the SPR linkages
- Assumptions, incomplete information, and uncertainty regarding dose–response behavior of receptors in the SPR linkage
- The limitations of toxicological science, especially with respect to the wide range of contaminants present in the environment, the understanding of cumulative effects, and the limited available information on ecological toxicology

Another important consideration is the likelihood that more than one SPR linkage exists at a given site. One source, for instance, may contribute contaminants that move through different pathways to different receptors. For example, a spill of volatile organic compounds (VOCs) may result in (1) DNAPL (dense nonaqueous phase liquid) density-driven migration along the top of a shallow bedrock horizon toward a river, (2) dissolved-phase contamination of groundwater being used for irrigation, and (3) vapor-phase transport in the unsaturated zone, which eventually contaminates shallow groundwater at some distance from the source point and in a different direction from (1) and (2).[2] In such a case, three separate SPR linkages exist, each of which has the potential to cause damage and needs to be accounted for separately. Each of the three risks will require separate estimates of damage costs.

Multiple SPR linkages also have implications for remedial decision making:

- The decision on the level of remediation required (setting the remedial objective) is complicated by having to consider the three linkages as separate issues, to some extent. In our example, for instance, the impact of DNAPL on the river may be ranked as the most urgent immediate problem, and a remedial objective might be to prevent DNAPL discharge to the river. However, dissolved-phase contamination, which could be migrating in a completely different direction (to irrigation wells), might be a longer-term problem, requiring a different objective.

- At one site, remedial decision making needs to consider three separate problems, to which the solutions may be quite different. In our example, the remedial techniques that apply to the DNAPL problem may not be effective for dissolved-phase contamination or vapor-phase migration.
- The technical feasibility of dealing with the different SPR linkages may also be vastly different. For example, remediating DNAPL, especially in deep, heterogeneous, or fractured systems, is extremely difficult, if not impossible, with present technology. A *remedial feasibility index* is introduced later, in Part III, to help account for this issue.
- The costs and benefits of dealing with each SPR linkage may have to be considered separately, in which case the decision-making process may involve a ranking of the risks, costs, and benefits of dealing with each separate problem.

Uncertainty will also be introduced into the decision-making process as a result of incomplete understanding of contaminant distribution, types, behavior, and mobility. This inherent uncertainty has a number of implications:

- Attempts to reduce technical uncertainty will generally involve increased data collection. The costs of data collection may be high, and there will clearly be a point of diminishing marginal returns in expenditure on data. This concept of *data worth* should be considered in any investigation. In the authors' experience, however, the point of diminishing marginal returns on data collection is rarely reached in practice. Many remedial decisions in the United States, the United Kingdom, and other countries are routinely made with insufficient data. The value of sufficient high-quality data cannot be underestimated. The results of risk assessment, the choice of remedial goal, and selection of remedial approach and technology are all based on the data collected at the outset. It is our experience that, in general, remedial activities cost at least one or more orders of magnitude more than data collection and review activities. A data worth analysis should be considered for each groundwater contamination case. Even if the analysis is basic and cursory, it will highlight the value of high-quality information to decision makers.
- Uncertainty can mean that groundwater contamination may not be detected for some time. The probability of detection of a given problem will tend to increase over time and with increased scale. The possibility that groundwater contamination is not detected and that damage results was discussed by Raucher in some detail (see Chapter 7).[3] This type of uncertainty should be considered in any cost–benefit analysis framework.
- Once detected, uncertainty may result in delays in determining the cause of contamination, the original source, and the responsible parties. Delays may result in additional costs and damages.

Any framework for incorporating cost–benefit analysis into remedial decision making for groundwater must include the ability to account for:

- Incorporation of the results of risk assessment, on which much of the current remedial guidance is based
- Existence of multiple SPR linkages on a given site
- Commingling of contaminant plumes, possibly involving several sources and several responsible parties
- Three fundamental modes by which groundwater-related risks may be generated
- Situations in which remedial objectives are set first, and then approach and technology options are evaluated to determine which reaches the objective most economically, or situations in which the costs and benefits of a range of fully developed remedial options are assessed to determine the remedial objective and the option that should be chosen
- Inherent uncertainties associated with predicting groundwater contaminant behavior and the associated risk

6.4 TIME AND SCALE

6.4.1 GROUNDWATER FLOWS

Groundwater flows, and contaminants move with it. Typical groundwater flow rates are in the order of centimeters or meters per year. Thus, the impacts of a groundwater contamination episode may not manifest themselves for several years or decades, reflecting the time it takes for the plume to migrate to the receptor. In the same way, as a plume migrates and spreads through an aquifer, impacting a larger and larger volume, the probability increases that more receptors will be affected. However, as time goes on, contaminant concentrations (and thus the risk-generating potential of the plume) may decrease, as a result of dilution, dispersion, adsorption, biodegradation, and chemical breakdown.

Groundwater contamination issues must be seen in the context of time and space and are inherently dynamic in nature. This presents a number of issues for setting remedial objectives and assessing the most economic remediation approach:

- Objectives must be framed in a temporal context — The level of risk associated with a given problem, and thus the predicted economic consequences should no action be taken, will change over time. In many cases, the longer we wait to deal with a problem, the worse it can get, and the more it may cost to deal with.
- Technology changes with time — The last 20 years have seen a significant amount of research into the detection, understanding, and remediation of groundwater contamination. What was considered technologically infeasible a decade ago may be wholly practicable and affordable today. This trend is bound to continue. In addition, the costs of remedial technologies

may change with time. For instance, the cost of air-strippers for removing VOCs from pumped groundwater has dropped significantly over the last ten years, while performance, ease of maintenance, and dependability have improved.

- Regulations change with time — In the United States, Europe, and the United Kingdom, the regulations dealing with groundwater contamination have been evolving for the last several decades. In Europe, new European Directives on water management and the environment are fundamentally changing the way such situations are reviewed, evaluated, and dealt with. Considering that planning horizons for serious groundwater contamination issues may be in the order of decades, or sometimes centuries (in the case of radioactive wastes, for instance), the likelihood is that relevant regulations and guidelines will change over the course of the project. Future changes at the European Union level (e.g., Water Framework Directive) will ensure continued evolution of the regulatory climate in Europe.
- Many deep groundwater contamination problems require long-term remedial solutions — In many cases, the only feasible remediation alternatives for groundwater contamination are containment or damage limitation, which involve long-term operation and maintenance of remedial systems. Pump-and-treat for plume control (hydraulic containment), for instance, is only effective while the pumps are running and the extracted groundwater is being treated at the surface. In cases where deep subsurface sources exist, pumping may have to continue indefinitely. Clearly, in these cases, time is a critical decision-making factor. Choosing an inappropriate planning horizon could compromise the decision-making process and result in selection of an infeasible and uneconomic remedial objective.

6.4.2 SCALE OF GROUNDWATER ISSUES

In the same way, the *scale* of a groundwater problem is not necessarily fixed. Groundwater contamination issues can vary in scale from plumes a few meters long to plumes covering several square kilometers. A spill that is initially concentrated in a small area may, over time, affect a considerable area, as groundwater carries the contaminants away and brings them into contact with other media and receptors. The scale of contamination may have significant impact on how the problem is valued by society:

- Larger-scale problems are likely to affect more people and a greater number of other receptors, all other things being equal.
- Larger-scale issues are more likely to involve a larger number of more diverse stakeholders, all of whom may wish to participate in decision making.

- Larger-scale issues are more likely to attract public attention, which may be reflected in media, public, and political scrutiny, and which may shift the economic and social perspectives of the stakeholders.
- Larger-scale problems are more likely to involve issues that transcend individual site decision making. Problems that cover large areas or cross jurisdictional boundaries may come to be seen as regional, or even national, in importance.

6.4.3 CUMULATIVE IMPACTS

Decisions on remedial objectives may, in some cases, need to reflect more than just the site-specific or problem-specific issues. For example, loss of any one aquifer may not be significant on a national or regional scale. Suitable, cost-effective alternatives may be available. In such cases, as argued by Raucher, economic analysis may reveal that remediation or restoration does not generate positive net benefits, and no action should be taken.[4] However, if this decision, taken in isolation, is repeated throughout the country or a particular region, the cumulative effect of the loss of several aquifers could be devastating. Thus, the scale of consideration of the problem is vitally important. This again reflects the need to consider the wider economic picture when setting remedial objectives.

Another major implication of time and scale issues for remedial decision making is that remedial objectives may change with time. As discussed in Chapter 2, the temporally variable nature of groundwater contamination problems may require a set of evolving remedial objectives that suit the conditions at the time. As regulations, public perceptions, technology, and global environmental conditions change, so too may remedial objectives. Any framework developed for groundwater remediation decision making should provide this type of flexibility. Clearly, a tiered system would be preferred, in which small, readily remediated problems can be assigned a relatively short planning horizon and a single-point remedial objective. Larger, more complex problems may require a more detailed analysis and definition of several remedial objectives over various planning horizons.

6.5 GROUNDWATER QUALITY AND QUANTITY

It is important to note that, in many situations, the quantity of available groundwater is just as important as its quality. Measures or actions designed to protect or remediate groundwater quality may also affect the quantity of groundwater available for use or as contribution to the hydrologic cycle. Examples of effects on groundwater quantity include:

- Pumping for remediation or containment, which lowers groundwater levels in an aquifer, reducing flows available to other users and affecting the water balance of surface water systems

- Placing restrictions on groundwater use to prevent inducing movement of contaminants toward wells or well-fields
- Damage to aquifer recharge zones, eliminating or reducing the effective recharge to an aquifer, and limiting the safe yields of groundwater for users

Therefore, economic analysis of remedial objectives should consider the possible effects on groundwater quantity, as well as quality.

6.6 IRREVERSIBILITY

6.6.1 IRREVERSIBLE AQUIFER DAMAGE

In the worst case, groundwater contamination may be irreversible for all practical purposes (within a few generations). Examples include situations involving deep contamination of highly heterogeneous media by nonaqueous phase liquids (such as hydrocarbons, organic solvents, and coal tar — see Chapter 3), radioactive contamination, and extensive contamination by compounds that tend to adsorb to the aquifer matrix material and are only released again slowly over time by diffusion. In these situations, little can be done in the near or medium term to reverse the damage and restore the affected aquifer. In most such cases, the best approach is to isolate the damaged area, contain the contaminants, and prevent them from affecting a greater volume of the aquifer. These may be termed *conditions of perpetual maintenance*, where, for the foreseeable future, an isolation or containment system will have to be operated, maintained, and monitored.

In such situations, the benefits of remediation may be clear, both on an intuitive level and based on a wider economic analysis. By definition, irreversible damage is beyond repair. However, care must be taken when using the term *irreversible*. In the final analysis, almost any subsurface contamination problem can be remedied if sufficient resources are put to the task. Even the examples listed previously could be remediated by excavating the subsurface sources, as would be done in an open-pit mine. Even then, the removed aquifer material would have to be replaced carefully with a substitute material. Clearly, the costs of such extreme solutions would be prohibitive. What is implicit in the term *irreversible* is an upper limit on society's willingness to pay for a solution. As discussed earlier, however, the future is uncertain. Should conditions change substantially or catastrophically, creating severe and life-threatening shortages of clean water, for instance, such irreversible damage could well be seen as reversible.

6.6.2 EXAMPLE: NAPLs IN FRACTURED AQUIFERS

The problems associated with NAPLs (nonaqueous phase liquids) in fractured rock and aquifers have received significant attention in the technical literature over the last decade.[5-7] Concerns over the impacts of chlorinated solvents on groundwater have led to a significant body of work examining DNAPLs in the subsurface, including in fractured rocks.[8,9] More recently, the unique behavior and problems associated with LNAPLs in fractured aquifers have been studied.[10–12] Chapter 3 provides a technical overview of NAPL behavior in the subsurface.

Remediation of LNAPL and DNAPL in fractured aquifers is a complex undertaking and, in many circumstances, may be considered practically unachievable. DNAPLs may migrate to significant depths via fractures, and if spill volumes are large and fracture interconnectivity high, DNAPL may invade progressively smaller aperture fractures with depth.[9] As NAPL fluid pressures increase, invasion of the rock matrix itself may also occur. The vertical migration of LNAPL in fractured aquifers is constrained by the water table, but despite this, significant penetration beneath the water table may occur, and lateral migration may occur in directions independent of the hydraulic gradient.[10] Within fractured aquifers, NAPL movement is governed by the geometry of the fracture network (including fracture orientations, densities, interconnectivity, apertures, and wall roughness), capillary pressure and fluid saturation relationships, and the properties of the NAPL (density, interfacial tension, viscosity).

Whether dealing with LNAPL, N-NAPL (neutral-buoyancy NAPL), or DNAPL, significant challenges exist when contemplating remediation. First, characterization of the distribution and behavior of NAPLs in fractured rock is notoriously difficult.[7,13,14] In a deterministic approach, fracture networks must be characterized, major fracture sets identified in the field, and representative fracture parameters determined. The occurrence of NAPL within these fractures then needs to be ascertained, areally and vertically. For DNAPLs, definitive characterization to depth may be problematic.[7,8,15,16] Rarely in practice is a complete characterization feasible.

Next, proven techniques for NAPL removal from fractures are few. Pump-and-treat methods, although effective for containment, have proved disappointing for NAPL removal, even when coupled with targeted NAPL recovery pumping and skimming.[17] Recently, more aggressive *in situ* NAPL-removal methods have been field tested, including high vacuum extraction, thermal heating,[18] and surfactant-assisted aquifer remediation.[19] These relatively expensive methods have showed good results in some cases but have not yet been rigorously tested in fractured rock environments. Finally, when the understanding of contaminant distribution is sketchy, even the simplest remediation techniques can prove unsuccessful. The combination of new or unproven remedial techniques, incomplete characterization, and complex aquifer and contaminant distribution conditions makes remediation success uncertain.

From this perspective, this type of groundwater contamination problem may be considered irreversible in the near term. Dependable techniques for NAPL removal from fractures, especially at depth, are not currently available. Experimental techniques for removing NAPL from fractures are expensive to apply, and success is far from assured. In some cases, problem holders and regulators have resorted to labeling such situations as beyond the current capabilities of technology. Several of the recent technical impractability (TI) waivers issed by the USEPA have been for fractured rock sites. Remedial decision making then concentrates on other, more feasible remedial objectives than source removal, such as protection of receptors through pathway management or removing or managing receptors through some form of institutional control on the affected groundwater resources. Here, a complete economic analysis can be useful in helping to identify and justify situations in which NAPL remediation would not be "economic". Thus, cost–benefit analysis could be used as a determining criterion for technological impractability (infeasibility): if the resources or receptor being damaged were valuable enough to society, sufficient funds could conceivably be brought to bear to remediate even the most difficult and complex NAPL spill.

Within this context, a clear understanding is required of the financial and broader economic implications of this type of problem. Available technical and scientific literature focuses on the application of specific remedial techniques and technologies to groundwater problems and deals almost entirely with remedial costs, cost comparisons, and cost effectiveness. The wider benefits of remediation are rarely discussed but include contamination damage avoided by remediation.[20,21] Much of this work is of primary interest to problem holders, but even so, very little is available that discusses the private benefits of remediation that accrue to problem holders. In Chapter 13, a case history involving NAPL contamination in fractured rock, among other problems, is considered in detail and a cost–benefit analysis completed.

REFERENCES

1. USEPA, Records of Decision, USEPA, Washington, DC, 1980.
2. Mendoza, C.A. and McAlary, T.A., Modeling of groundwater contamination caused by organic solvent vapors, *Ground Water*, 28, 2, 199, 1990.
3. Raucher, R.L., A conceptual framework for measuring the benefits of groundwater protection, *Water Resour. Res.*, 19, 2, 320, 1983.
4. Raucher, R.L., The benefits and costs of policies related to groundwater contamination, *Land Econ.*, 62, 1, 33, 1986.
5. Mackay, D.M. and Cherry, J.A., Groundwater contamination: pump-and-treat remediation, *Environ. Sci. Tech.*, 23, 6, 630, 1989.
6. Mercer, J.W. and Cohen, R.M., A review of immiscible fluids in the subsurface: properties, models, characterization and remediation, *J. Contaminant Hydrology*, 6, 107, 1990.
7. Pankow, J.F. and Cherry, J.A., Eds., *Dense Chlorinated Solvents and Other DNAPLs in Groundwater*, Waterloo Press, Portland, OR,1996.
8. Cohen, R.M. and Mercer, J.W., *DNAPL Site Investigation*. C.K. Smoley, Boca Raton, FL, 1993.
9. Kueper, B.H. and McWhorter, D.B., The behavior of dense nonaqueous phase liquids in fractured clay and rock, *Ground Water*, 29, 5, 716, 1991.
10. Hardisty, P.E., Wheater, H.S., Johnson, P.M., and Bracken, R.A., Behavior of light immiscible liquid contaminants in fractured aquifers, *Geotechnique*, 48, 6, 747, 1998.
11. Wealthall G.P., Kueper B.H., and Lerner D.N, Fractured rock-mass characterization for predicting the fate of DNAPLS, in *Conf. Proc. Fractured Rock 2001*, Kueper, B.H., Novakowski, K.S. and Reynolds, D.A, Eds., Toronto, 2001.
12. Hardisty, P.E., Wheater, H.S., Birks, D., and Dottridge, J. Characterization of LNAPL in fractured rock, *Q. J. Eng. Geol. Hydrogeol.*, 36, 343, 2003.

13. CL:AIRE *Introduction to an Integrated Approach to the Investigation of Fractured Rock Aquifers Contaminated with Non-Aqueous Phase Liquids*. Technical Bulletin TB1, London, 2002.

14. Hardisty, P.E., Johnston, P.M., Honhon, P., and Young, R. Characterization of occurrence and behaviour of LNAPL in fractured rocks, *Proc. API/NGWA Conf. Hydrocarbons and Org. Chemicals in Groundwater*, Houston, 1993.

15. Guswa, J.H., Benjamin, A.E., Bridge, J.R., Schewing, L.E., Tallon, C.D., Wills, J.H., Yates, C.C., and LaPoint, E.K. Use of FLUTe systems for characterization of groundwater contamination in fractured bedrock, *Fractured Rock 2001 Conf. Proc.,* Kueper, B.H., Novakowski, K.S., and Reynolds, D.A., Eds., Toronto, 2001.

16. Lane, J.W., Jr., Buursink, M.L., Haeni, F.P., and Versteeg, R.J., Evaluation of ground penetrating radar to detect free-phase hydrocarbon in fractured rocks: results of numerical modelling and physical experiments, *Ground Water,* 38, 6, 929, 2000.

17. Schmelling, S.G. and Ross, R.R., *Contaminant Transport in Fractured Media: Models for Decision Makers*, EPA Superfund Issue Paper EPA/540/4-89/004, 1989.

18. USEPA, Steam injection combined with electrical resistance heating at the Young-Rainey STAR Center, *Tech. News and Trends*, 10, 2004.

19. Taylor, T.P., Pennell, K.D., Abriola, L.M., and Dane, J.H., Surfactant enhanced recovery of tetrachloroethylene from a porous medium containing low permeability lenses: 1. Experimental studies, *J. Contaminant Hydrology*, 48, 325, 2001.

20. Peramaki, M.P. and Donovan, P.B., A remedial-goal driven, media-specific method for conducting a cost benefit analysis of multiple remedial approaches, *Proc. NGWA Conf. on Remediation: Site Closure and the Total Cost of Cleanup*, New Orleans, November, 2003.

21. Goist, T.O. and Richardson, T.C., Optimizing cost reductions while achieving long term remediation effectiveness at active and closed wood preserving facilities. *Proc. NGWA Conf. on Remediation: Site Closure and the Total Cost of Cleanup*, New Orleans, November, 2003.

7 Economic Theory for Groundwater Remediation

Chapter 4 reviews the literature that deals specifically with valuation of groundwater in its own right, and Chapter 5 introduces the framework for economic analysis in general. Chapter 6 outlines the specific economic issues associated with groundwater contamination and remediation. This chapter brings these three discussions together, reviewing the research that attempts to value the benefits of groundwater protection or remediation and setting the theoretical background for the operational framework developed in Part III. The amount of research on this issue cannot yet be described as substantial, although it is clear that considerable effort has been put into the subject in the last few years. The need for this type of work has been expressed by several authors, including the National Research Council in the U.S., which states that what is most relevant for groundwater pollution policy decision making is knowledge about how economic values are affected by the implementation of those decisions.[1]

7.1 CONCEPTUAL FRAMEWORKS FOR BENEFITS ASSESSMENT

Raucher, in a landmark paper, describes a conceptual framework for measuring the benefits of groundwater protection.[2] The discussion focuses on benefits, concluding that the costs and feasibility of various groundwater protection measures were (even at that time) relatively well understood. The framework presented is best expressed in Equation 7.1:

$$E(NB_i) = E(B_i) - X_i \qquad (7.1)$$

where $E(NB_i)$ is the expected net benefits of an activity that would enhance groundwater protection, $E(B_i)$ denotes the expected social benefits of groundwater protection strategy i, and X_i denotes the social costs associated with implementation of that strategy. Raucher approximated the social costs X_i by the cost of executing the protection measure (relatively well understood).[2] However, as Chapter 8 shows, external costs (or secondary impacts of remediation) should be added to this definition of costs.

The benefits of groundwater protection, on the other hand, are defined by the change in expected damage, E(D), as in Equation 7.2:[2]

$$E(D) = p\left[qC^r + (1-q)C^p\right] \tag{7.2}$$

where p is the probability, in the absence of strategy i, that contamination will occur; q is the probability that groundwater contamination would be detected before contaminated water was used; C^r is the cost of the most economically efficient response to the problem; and C^u is the cost incurred if contaminated water continues to be used in the same way as prior to the incident (or *damage*, to use the terminology of Part II).

Raucher uses this framework to put forward some very powerful arguments.[2] First, the inclusion of a probabilistic component to the framework explicitly recognizes the role of uncertainty in the decision-making process. He notes that both p and q are highly dependent on a sound understanding of the hydrogeological regime and the behavior of the contaminants within the groundwater system. The role of predictive modeling would be key in applying this framework in practice. Much information is available on the uncertainties and limitations of groundwater flow and contaminant fate and transport modeling.[3-5] Also, the literature dealing with quantitative risk assessment is replete with discussions of the limitations and uncertainties inherent in assessing the risks posed by groundwater contamination.[6]

The possibility that groundwater contamination may not be detected for a considerable period, during which it migrates over potentially large distances and affects a greater number of receptors, is a key issue. This highlights the value of investment in monitoring and investigation programs. Raucher argues that the probability of detecting contamination q is not fixed in time and will increase as the plume moves and affects a greater area, and also if policies are put into place that are likely to improve the possibility of detection.[2] The framework can thus be used to estimate the net benefits expected from the implementation of a detection policy.

The endpoints of the argument reveal some interesting assertions. If, in the worst case, contamination is certain ($p = 1$) but impossible to detect ($q = 0$), groundwater would continue to be used in the same way and damage would result. The expected damage $E(D) = C^u$. Alternatively, if contamination remained certain ($p = 1$), but there was full certainty of detection ($q = 1$), then the least-cost remedial solution would be implemented, and so $E(D) = C^r$. Hence, for a monitoring policy that improved q, the expected net benefit would be $C^u - C^r$.

Second, Raucher's framework brings forward the possibility that economic analysis might reveal conditions where C^r is greater than or equal to C^u, in which case it may be better, from an economic standpoint, to do nothing.[2] This simple model shows how the setting of a remedial objective can be framed in economic terms. This subject is a key focus of this book and will be discussed in detail in Part III. Definition of the costs of contamination is strongly site specific. In considering C^u, factors such as site hydrogeology; contaminant type, toxicity, and behavior in the aquifer; and the use of the water are important. Response costs (C^r) should be based on the least cost of all feasible options that will prevent or remediate the contamination. However, we note that this assumes full information and rationality of decision making and does not explicitly include the uncertainty associated with remedial system performance and the likelihood that aquifer remediation may lead to only a *partial* clean-up.

Raucher groups response options into three categories:[2]

- Restoration
- Containment
- Avoidance

Restoration is not considered in Raucher's paper, because it is deemed technologically infeasible.[2] Containment refers to control of the spread of contamination and includes pump-and-treat and physical barriers. Avoidance options deal with treatment at point of use or development of alternative sources of water.

Raucher presents an important discussion on the issue of time, as it applies to his framework.[2] The author found that the choice of planning horizon had a significant impact on the cost–benefit calculations. Over time, growth of a plume would tend to affect a greater number of receptors, resulting in increased damage costs. Longer planning horizons allow for accrual of greater total benefits, all other factors remaining equal, simply by virtue of summing benefits over a greater number of years. More subtle time–benefit relationships may also exist. Over a longer period of time, for instance, there may be increased likelihood of contamination occurring (e.g., due to waste container degradation).

Nonuse benefits (see Chapter 4 for a definition and examples) are also discussed by Raucher, who points out that his benefits-formulation framework does not incorporate a premium accounting for option value (the WTP to ensure access to a resource at some point in the future, regardless of whether it is currently being used).[2] This may lead to a significant understatement of the benefits of protection, especially when the potentially irreversible nature of groundwater contamination is considered.[7] Raucher, however, presents two alternative formulations in an appendix, designed to allow incorporation of nonuse values into the framework.[2] One option involves direct incorporation of nonuse benefits as a separate term from the use values. Another involves addressing the nonuse values through extending the time horizon indefinitely or reducing the discount rate to zero.

Finally, Raucher presents a series of hypothetical examples, illustrating the workings of the framework.[2] The results show that altering the discount rate and water use for two types of groundwater plume (large and slow, and small and fast) changes the response deemed to be most economic. These findings reinforce the broad conclusions of the work: that economic analysis of the benefits of groundwater protection and remediation is site specific; that the calculated best response is highly sensitive to time and the discount rate; and that the uncertainties associated with the lack of full hydrogeological knowledge, risk assessment limitations, and the lack of research into nonuse groundwater values make practical application of this type of framework quite difficult.

In more empirical research, Kulshreshtha presents one of the few detailed and comprehensive analyses of the economic valuation of groundwater, examining the Assiniboine Delta aquifer in Manitoba, Canada.[8] He estimates the total economic value of groundwater in the aquifer by considering all use and nonuse values. Direct uses include irrigation, domestic supply, and industrial and thermal uses of ground-water. Indirect uses include recreation and tourism (in the form of groundwater contribution to nearby lakes and streams) and environmental use. Option values are also considered, as are existence value and bequest value. Thus, annualized net present values for each use category can be summed over a given period of time to produce an estimate of total economic value, as in Equation 7.3 and Equation 7.4:

$$NPW_w = \sum_{i=1}^{s} NPVW_i \qquad (7.3)$$

$$NPVW_i = \sum_{t=1}^{L} \frac{VW_{it}}{(1+r)^t} \qquad (7.4)$$

where NPW_w is the net present worth of the aquifer, $NPVW_i$ is the net present value of water in the ith use ($i = 1....s$), VW_{it} is the total value of water in the ith use for the year t ($t = 1...L$), and r is the discount rate. Numerical results of this research are presented in Chapter 4, in Table 4.2.

The discussion in the remainder of this chapter is divided into two main parts, addressing two levels of uncertainty:

- When contamination and damage are known to be occurring
- When contamination is not suspected or is unknown, and damage is occurring without our knowledge

Although it is possible to complicate the theory very rapidly, especially when considering the case where contamination is not suspected, the theoretical framework presented here has been kept as straightforward as possible, because the ultimate goal is to develop an accessible, easy-to-use framework. For more detailed and technical treatment of uncertainty in the context of groundwater, the reader is invited to see Edwards, Huan et al., and Forsyth for quality-related issues, and Tsur et al. and Rubio et al. for quantity-related issues.[9-13]

The theoretical economic framework presented earlier provides a broad direction for development of a decision-making support tool based on economic analysis. However, as will become apparent from the following discussion, at its most complex, full analysis of costs and benefits can be a daunting task. The elements of spatial and temporal variability in benefit, damage, and cost functions, together with the ability to take remedial (corrective) and preventative (avoidance) actions at different times and points in space, provide for a complex analytical problem. However, it is very likely that, in many situations, a relatively simple and straightforward analysis will be all that is warranted (or all that is possible, considering constraints of time and resources). Only for the most serious and difficult problems will a fully comprehensive analysis be desired. The analysis framework presented in Part III is designed with this in mind.

7.2   BENEFITS AND COSTS OF GROUNDWATER REMEDIATION WHEN CONTAMINATION AND DAMAGE ARE KNOWN

In order to keep the methodology simple, we assume that there are just two actions we can take when groundwater contamination is known:

- Avoidance — Prevention of further damage, for which we use the notation A
- Remediation of existing damage — For which we use the notation R

Each action can be taken at different points in time, so there is, in fact, a substantial array of policy options. For example, preventive action could be taken now, thus avoiding future damage, or action can be postponed, suffering existing damages now and future damage at the current level (or higher damages if contamination gets worse, or lower damages if contamination gets better). For each action, the other variable is spatial — the location at which the avoidance or remediation takes place. For instance, a barrier or containment system could be located close to the source, capturing the most concentrated contaminants and allowing the remainder to escape. Or, if placed further down-gradient, the containment system could capture all of the known contamination. A third option is to combine avoidance and remedial actions (AR).

7.2.1   BENEFITS — PREVENTION ONLY

Remedial objective setting must consider the benefits of the action. If the overarching aim is to maximize human welfare (which implicitly includes protection of the environment and ecosystems on which human life depends) in the context of present value, then we would select the action and time-phasing that maximize the net present value (NPV) of benefits and costs.

Let baseline damages be D_t (i.e., damages that occur when *no* remediation action is taken). Typically, damage increases over time (or $dD/dt > 0$) because the plume migrates over time, potentially impacting a greater number of receptors and reducing the groundwater resource base. D_t can in fact be a complex function, depending on the nature of the contaminant, the speed of movement, the assets at risk, and the economic value of those assets. For instance, should plume migration be coincident with significant contaminant attenuation (through dilution, dispersion, adsorption, and chemical and biological breakdown), the severity of the impact to receptors may decrease with time, despite more receptors' being affected. In this case, the function D_t will likely have a convex-up character, with a steep initial slope, a maximum ($dD/dt = 0$), and a long tail. For continuous sources, damage could extend over long periods of time. For instantaneous or slug sources or events, damages may tend to decrease with time. These concepts are illustrated schematically in Figure 7.1, using a simple construction of benefits–costs–damages vs. time. The function Dt in Figure 7.1 is an increasing damage function.

If it is assumed that the only action is prevention, and if action is taken in time 0 ($t = 0$), the path of damages is given by D_{A0} in Figure 7.1 (i.e., damage contingent

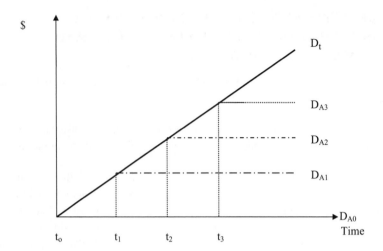

FIGURE 7.1 Benefits–damage–cost vs. time: avoidance.

on acting in period 0). If action is taken at time 1 ($t = 1$), the result is D_{A1}, and so on. In all cases, D_{An} starts at 0. The exact shape of D_{An} will depend on:

- Nature of the plume
- Nature of the contaminant
- Any attenuation, natural or otherwise, experienced by the contaminant

Thus, D_{A1} might appear as any of the curves in Figure 7.2.

The benefits of taking action $_{An}$ are derived as shown in Equation 7.5:

$$B_{An} = \sum_{n}^{T} D_t \qquad (7.5)$$

where T is the time horizon (planning horizon, or the length of time over which the situation is to be considered) and D_t is baseline damage in year t. Thus, acting immediately ($t = 0$) avoids all the baseline damages over the planning horizon. If we act in any other period, the benefits equal baseline damages minus the damage incurred up to that period. Note that the conditional damages D_{At} comprise damage incurred from contamination in $t = 1,2,...T$, plus past damage in $t = 0,1,...T-1$. The shaded area in Figure 7.3 represents the benefits of acting in period $t = 1$. This is represented mathematically in Equation 7.6. The benefit of action taken at period t (B_{At}) is equal to the sum of the differences between damages that would have occurred through time (Dt) and the damages that result if avoidance is taken at time t (see Equation 7.6).

$$B_{At} = \sum_{0}^{T} D_t - D_{At} \qquad (7.6)$$

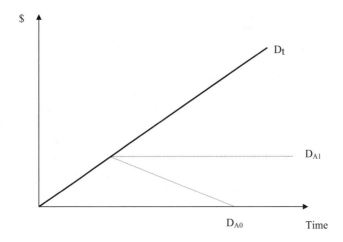

FIGURE 7.2 Alternative avoided damage curves.

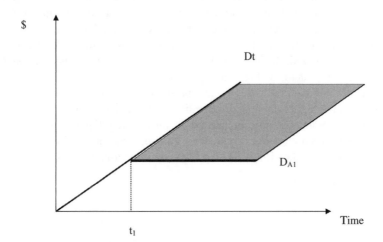

FIGURE 7.3 Benefits of action taken at time $T = T_1$.

7.2.2 BENEFITS — PREVENTION AND REMEDIATION

Remediation is now introduced to the preceding analysis (shown as *AR*). When remediation takes place, benefits become all damages over time (avoided by *AR*) minus the damage incurred up to the point in time when the combined prevention and remediation action (*AR*) is taken.

For illustration, say the remediation action takes place in period 2 such that further damage is prevented. The benefits are described by Equation 7.7:

$$B_{AR2} = \sum_0^T D_t - D_{AR2} \qquad (7.7)$$

The benefits of this intervention (AR) are given by the shaded area in Figure 7.4. The area represented by the small triangle below Dt, from t_0 to t_2, represents the value of the damage that occurred before the combined remedial and prevention action was taken. In this case, the action has prevented all future damages and would represent a situation in which the impacted portion of an aquifer was completely restored and the ongoing migration of contamination to new receptors was halted.

A more typical case would involve partial restoration of an aquifer and complete or partial containment of future damages. In this case, the D_{AR} curve would be nonzero, and given that the restoration action would take several years, the curve would probably have a negative slope, steep at the beginning and reaching an asymptotic nonzero value over time. This behavior is typical for aquifer restoration projects. Such a curve is shown in Figure 7.5.

7.2.3 NET BENEFITS

What is important in a decision-making process is not only the magnitude of benefits but also the magnitude of costs that will bring about the said benefits — hence the term *cost–benefit* analysis. For simplicity, assume that *prevention costs* (C^p) are the same regardless of *when* we intervene (i.e., $C_t^P = C_{t+1}^P$ and so on). This implies that it costs the same (in current-year prices) to adopt a preventive policy regardless of when we act. In practice, C^p may vary, depending on the nature of the compounds and processes involved, the hydrogeology of the site, and the industry. *Remediation costs*, on the other hand, will tend to vary with size of plume, type of contaminant, and the nature of the aquifer material and properties. In at least some cases, the later we intervene, the higher the cost of remediation (C^r) will be. This case is illustrated in Figure 7.6. Note that care is needed in interpreting Figure 7.6. The cost of prevention C^p does not occur every year — it simply says prevention cost equals C^p given that action is taken that year, (i.e., the conditional notation C_{At}^P).

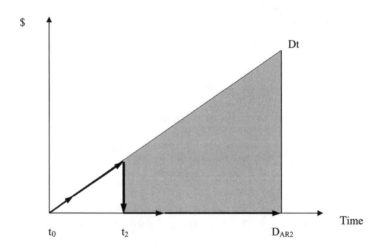

FIGURE 7.4 Benefits of prevention and remediation (full restoration = ideal case).

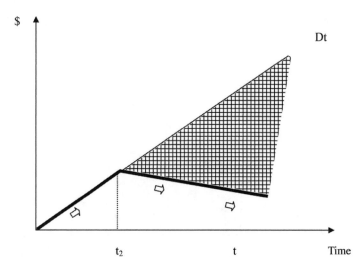

FIGURE 7.5 Benefits of prevention and remediation (typical case).

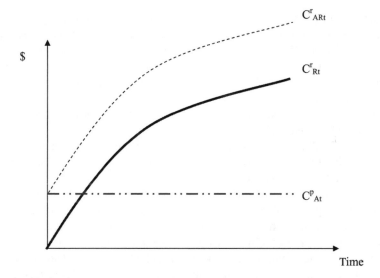

FIGURE 7.6 Costs of prevention and remediation.

To find net benefits, we deduct the flow of costs from the flow of benefits. As an example, the net (undiscounted) benefits of prevention minus the costs of the selected remediation policy in any year, t, is given by Equation 7.8:

$$\sum B_{ARt} - \sum C_{ARt} = \sum_{0}^{T} \left[D_t - D_{ARt} - C_{ARt}^P - C_{ARt}^r \right] \qquad (7.8)$$

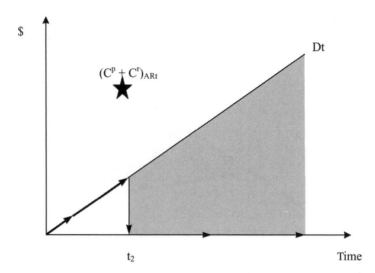

FIGURE 7.7 Net benefit of a prevention–remediation policy flow of benefits — flow of costs.

This is shown schematically in Figure 7.7. The arrows show the progression of damages over time as the policy is implemented, resulting in the benefits represented by the shaded area. This area represents an annual benefit accrued over a number of years, providing a total benefit value. In Figure 7.7, the costs of prevention and remediation are represented by the point $(C^p+C^r)_{ARt}$. This point represents the one-time cost incurred to produce the benefits shown in the shaded area. This example, of course, assumes that a single-year prevention and remediation intervention solves the problem instantaneously.

In practice, C^p and C^r will be distributed over time, as shown in Figure 7.8. More typically, especially in cases involving organic contamination of heterogeneous aquifers, remedial costs will tend to fluctuate considerably over time (Figure 7.9). An initial remedial system will be put into place, involving capital expenditure and operated over a period of some years (operating and maintenance costs incurred). At some point in time, the system will need capital upgrade and modification to deal with the changing conditions within the aquifer (the remediation is inherently acting to change the conditions and chemistry within the aquifer) and to account for the life span of key system components. This results in a characteristic spiked cost–time curve. Chapter 8 provides further discussion on costs of remediation.

7.2.4 DISCOUNTING

For simplicity, the preceding analysis was presented for undiscounted flows. When discounting is introduced, the net benefit equation (Equation 7.9) for prevention and remediation intervention becomes:

$$NPV_{ARt} = \sum_{0}^{T}\left[\frac{D_t - D_{ARt} - C_{ARt}}{(1+r)^t}\right] \tag{7.9}$$

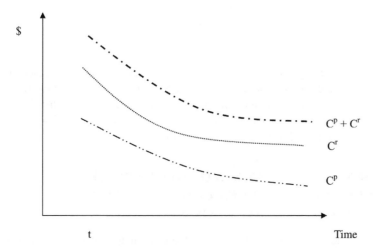

FIGURE 7.8 Costs of prevention and remediation as a function of time.

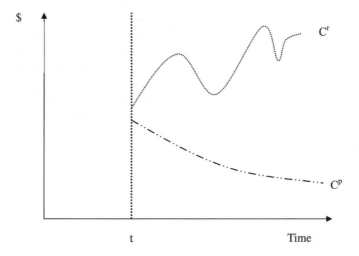

FIGURE 7.9 Typical life cycle remedial cost function.

where NPV is net present value of the net benefits over time, NPV_{ARt} is NPV conditional upon the A and R mix of intervention that takes place, C is the aggregate cost of prevention and remediation ($C^p + C^r$), and r is the discount rate. Note that the solution to the policy issue is given when NPV is maximized. This can be solved iteratively using available computer numerical algorithms, but only for relatively simple cost and benefit functions. Clearly, the type of complex, highly time-variant cost and benefit functions discussed previously will make mathematical optimization more difficult.

The optimum intervention time t^* is given when NPV approaches zero, as in Equation 7.10:

$$\frac{dNPV}{dt} = 0 \qquad\qquad (7.10)$$

Note that damage is likely to be probabilistic. So D should be replaced by some distribution function $f_d(D)$ and similarly with costs, $f_c(C)$. In practice, these distributions are likely to be known only as ranges of values. Simple distributions may have to be imposed on ranges. This is the basis for the *benefits threshold* concept used in the framework procedure, discussed in Chapter 9.

7.3   BENEFITS AND COSTS OF GROUNDWATER REMEDIATION WHEN CONTAMINATION AND DAMAGE ARE NOT KNOWN

Section 7.2 deals with the context in which contamination damage to the aquifer has already been detected. This section deals with the case where damage is not known. In other words, contaminants are moving within an aquifer and have reached one or more receptors without the knowledge of any individual or organization. An example of this could include contaminants discharging from an aquifer into a relatively remote and unmonitored river or wetland, where neither the groundwater contamination nor the resulting surface-water contamination has been detected or is suspected. Humans using the resource may also be unknowing receptors, and damage may be occurring over time. Another example, of a type that has been documented in the United States, is that of farmers using well water for irrigation and watering of animals. Without their knowledge, the aquifer and their well water can become contaminated resulting, over the years, in crop damage and animal mortality. Farr described a situation in Woburn, Massachusetts, where residents of a small town were unknowingly exposed to carcinogenic compounds in well water over a period of several years, resulting in widespread illness and the deaths of several children and adults in the community.[14]

This case differs from the analysis under certainty (when contamination is known) in two ways:

- An additional action — monitoring — is introduced
- There is uncertainty about whether contamination exists at all

7.3.1   MONITORING

Monitoring for contaminants can, in principle, apply to all aquifers. In reality, some aquifers will not be monitored if there is no reason to suspect that contaminants have affected or will affect them. Monitoring involves costs, C_t^m, which will occur over the lifetime of the monitoring system, but with some up-front capital costs of installing the monitoring system. In any year t, then, there will be costs of C_t^m for those aquifers that are monitored.

A key issue is whether monitoring is worthwhile. Are the costs of monitoring justified by the potential benefits of avoiding damage? Monitoring may take place and no damage be detected. Let the probability of detection (p_d) be 1 if contamination occurs (i.e., some state-of-the-art monitoring exists with 100% accuracy of detection). Let the probability of contamination (p_c) be $0 < p_c < 1$. Then what matters is the flow of net benefits that occurs with monitoring, as opposed to the flow without monitoring.

Without monitoring, damage will be detected only when it has occurred. Monitoring enables early action. We can suppose that early action is cheaper than action arising after damage has occurred. Hence, monitoring saves the difference between action costs without monitoring and action costs with monitoring, call it $\Delta(C^p + C^r)$. It will also avoid residual damages, d, that occur when damage is revealed, rather than being detected early. This, of course, assumes that the monitoring system is successful at detecting the contamination. Thus, the NPV of monitoring is the sum of avoided intervention costs and residual damage. Note that this assumes that action is immediate on discovering the damage. This can be expressed as in Equation 7.11:

$$E(NPV^m) = \sum_0^T \frac{\left\{ p_c \cdot \left[\Delta(C_t^p + C_t^r) + d \right] \right\} - C_t^m}{(1+r)^t} \qquad (7.11)$$

where $E(NPV^m)$ is expected value of the net present value of monitoring and other notation is as described previously.

NPV of monitoring here is contingent on the benefits of action exceeding the costs of action. Further adjustment could take place by assuming that the monitoring is not wholly accurate $(0 < p_d < 1)$. Then Equation 7.12 applies:

$$E(NPV^m) = \sum_0^T \frac{p_c \left\{ p_d \cdot \left[\Delta(C^p + C^r) + (1 - p_d) \cdot d \right] \right\} - C^m}{(1+r)^t} \qquad (7.12)$$

This is similar to what was covered in Raucher.[15] But whereas his formulation covers the net benefits of any policy intervention, our concern here is the benefits of monitoring only. Monitoring benefits accrue as differentials arising from the flows of benefits and costs in the event of contamination and detection, compared to what would happen if there were no monitoring (the baseline condition in this case).

7.3.2 Benefits with Uncertainty — Expected Utility

The preceding approach assumes that damage and control costs are treated in a risk-neutral way. This means that society values a unit gained in the same way it values the same unit lost. In practice, however, there is some degree of risk aversion in society, which implies that the unit lost is valued more than the unit gained. The importance of risk aversion is that the maximum sum that society will be willing to pay to secure monitoring will be different from the expected value of the damages and control costs.

The theoretically correct solution to the value of a monitoring system in the face of uncertainty involves the notion of *option price*. Explanation of the concept is

TABLE 7.1

An Illustrative Example for the Expected Value and Expected Utility Concepts

Contingency	Monitor	Do Not Monitor	Probability of Contamination
No contamination: Y_{-d}	110	100	$0.8\ (1 - p_c)$
Contamination: Y_d	100	50	$0.2\ (p_c)$
Expected value: E	108	90	
Variance	16	400	
Contingent surplus:			
No contamination: S_{-d}	+10		
Contamination: S_d	+50		
Expected value of surplus:	+18		
$E(CS)$			

somewhat involved, so it is illustrated here by an example.[16] Table 7.1 assumes that monitoring produces differences in income flows, where *income* here conflates benefits and costs. Hypothetical numbers are shown. Actions are monitor and not monitor. The contingent events are contamination and no contamination, with probability of contamination (p_c) set at 0.2 for illustration. Table 7.1 shows that the expected value of monitoring benefit flows is 108, compared to 90 with no monitoring.

Expected values are simply calculated as in Equation 7.13:

$$E = p_c \cdot Y_d + (1 - p_c) \cdot Y_{-d} \qquad (7.13)$$

where Y_{-d} and Y_d are the payoffs (or economic value of aquifer) in no-contamination and contamination cases, respectively. Thus, 108 is equal to $(0.8 \times 110 + 0.2 \times 100)$.

The variance is shown to illustrate the wide range of potential outcomes in the no-monitoring case (i.e., uncertainty is substantial). Variance of the expected value is calculated as in Equation 7.14:

$$E = p_c(Y_d - E)^2 + (1 - p_c)(Y_{-d} - E)^2 \qquad (7.14)$$

The contingent surplus is derived by reading across the table. In Table 7.1, for the no-contamination case, the WTP for monitoring is $110 - 100 = 10$.

The expected value of the surplus, $E\ (CS)$, is p_c, and $S_d + (1-p_c)\ S_{-d}$. This is the "normal" measure of the value of the monitoring system.

Option price is the maximum WTP of individuals (society) for a policy in a context in which the benefits of the policy are uncertain. So, EU becomes the following, and the option price for the policy can be found from Equation 7.15:

$$EU_m = p_c \cdot U(Y_d - OP) + (1 - p_c) \cdot U(Y_{-d} - OP) = EU_{-m} \qquad (7.15)$$

where Y is the payoff, U is the utility attached to the payoff, OP is option price, and the notations d and $-d$ mean contamination and no-contamination, respectively, in the with-monitoring case. The notations $-m$ and m mean the no-monitoring case and monitoring cases, respectively. In the preceding example, this would be:

$$EU_{-m} = [0.2.U \times (110 - OP)] + [0.8.U \times (100 - OP)]$$

First, we find EU_{-m}, which is the expected utility without the monitoring policy. Read the payoffs from the Do Not Monitor column, and EU is given by $(0.8U \times 100 - 0.2U \times 50)$ or by Equation 7.16:

$$(1 - p_c) \cdot U(Y_{-d}) - p_c \cdot U(Y_d) \tag{7.16}$$

To simplify, we can assume that $U(Y) = lnY$. The EU_{-m} is then 3. Substitute this in the previous equation containing OP to obtain:

$$0.2.\ln(110 - OP) + 0.8.\ln(100 - OP) = 3$$

This can be solved iteratively (insert preliminary estimates of OP and iterate) to obtain $OP = 45$. Compare OP and E (CS), and we see that OP is substantially larger. OP then needs to be compared with the costs of the policy, and OP greater than cost would be the requirement to proceed. Note that if costs are between 19 and 44, the policy would be adopted, but it would not have been adopted under the conventional rule that E (CS) should be greater than the costs.

This example illustrates the complexity of including an explicit accounting for risk adversity within a guidance framework. Because a step to evaluate risk aversion is not thought to be relevant for most applications, it is not explicitly included in the framework presented in Part III.

REFERENCES

1. National Research Council, *Valuing Groundwater*, NRC, Washington, DC, 1997.
2. Raucher, R.L., A conceptual framework for measuring the benefits of groundwater protection, *Water Resour. Res.*, 19, 2, 320, 1983.
3. Freeze, R.A. and Cherry, J.A., *Groundwater*, Prentice Hall, Upper Saddle River, NJ, 1979.
4. Bear, J. and Verruijt, A., *Modeling Groundwater Flow and Pollution*, D. Riedel, Dordrecht, 1987.
5. Fetter, C.W., *Contaminant Hydrogeology*, Macmillan, New York, 1992.
6. ASTM, *Risk-Based Corrective Action Applied at Petroleum Release Site,*. ASTM Standard Guide E1739, 1995.
7. Kavanagh, M. and Wolcott, R.M., *Economically Efficient Strategies for Preserving Groundwater Quality*, prepared for the USEPA by Public Interest Economics, Washington, DC, 1982.

8. Kulshreshtha, S.N., *Economic Value of Groundwater in the Assiniboine Delta Aquifer in Manitoba*, Social Science Series, 29, Environmental Conservation Service, Environment Ottawa, Canada , 1994.
9. Edwards, S.F., Option prices for ground water protection, *J. Env. Econ. Manage.*, 15, 474, 1988.
10. Huan, W.Y. and Uri, N.D., An analytical framework for assessing the benefits and costs of policies related to protecting groundwater quality, *Environment and Planning A*, 22, 1469, 1990.
11. Forsyth, M., The economics of site investigation for groundwater protection sequential decision making under uncertainty, *J. Environ. Econ. Manage.*, 34, 1, 1, 1997.
12. Tsur, Y. and Zemel A., Uncertainty and irreversibility in groundwater resource management, *J. Environ. Econ. Manage.*, 29, 149, 1995.
13. Rubio, S.J. and Castro J.P., Long-run groundwater reserves under certainty, *Investigaciones Economicas*, 20, 1, 71, 1996.
14. Farr, J., *A Civil Action*, Prentice Hall, Upper Saddle River, NJ, 1996.
15. Raucher, R.L., The benefits and costs of policies related to groundwater contamination, *Land Econ.*, 62, 1, 33, 1986.
16. Hardisty, P.E. and Ozdemiroglu, E., *Costs and Benefits Associated with Remediation of Contaminated Groundwater: A Framework for Assessment*, U.K. Environment Agency Technical Report P279, Bristol, 2000.

8 Remedial Costs

This chapter provides an overview of the costs of remediation — both financial (or private) and external (or social) costs.

8.1 BACKGROUND: REMEDIAL TECHNOLOGY COSTS AND COST COMPARISONS

8.1.1 SEMANTICS

The groundwater literature contains many studies that compare the costs and "benefits" of two or more remedial methods or technologies for a particular contamination problem at a particular site. The word *benefits* in the previous sentence has been placed in quotation marks because, in the vast majority of cases, it is used incorrectly from an economist's point of view. Benefits, in the lexicon of the economist, were described in detail in Chapter 4. The term refers to specific improvements in the welfare of private and external stakeholders, measured in monetary units. The benefits usually being considered in the technical groundwater literature are, in fact, better described as *advantages* or *positives* and are usually expressed qualitatively, not in units of money. Examples of positive attributes commonly termed "benefits" in the groundwater literature are listed in Table 8.1. These should be included in a discussion of costs and are thus included in this chapter.

As an example, James et al. (1996)[1] compared the cost effectiveness of two remediation alternatives (containment and monitoring only) for radioactive waste affecting groundwater at the Oak Ridge National Laboratory in the United States.[1] They include the concept of *data-worth analysis*, to estimate maximum justifiable expenditures on data collection. Despite the title "Allocation of environmental remediation funds using economic risk-cost-benefit analysis: a case study," the discussion does *not* consider the benefits of the remediation, as the term is understood by economists and as it is used in this book. Rather, the study is one of cost-effectiveness analysis (CEA), with the term "benefits" actually used in the context of the ability of the technique to provide a certain level of remediation or to reach a predefined remedial goal. As discussed earlier, in most of the groundwater literature, the term *benefits* has been taken in a very narrow and limited context, primarily associated with the ability to achieve a set remedial target. These studies are more accurately described as CEAs or simple cost-comparisons, and pertain to the second overall objective of determining the most cost-effective remedial solution to achieve the objective. This confusion between the term *benefits* as used in the technical literature and the economic literature is widespread.

TABLE 8.1

Advantages Frequently Described as Benefits in the Literature

Remediation Advantage Described as a "Benefit"	Description
Fast, timely	Achieves desired concentration target relatively quickly compared to other methods
Effective	Removes a large mass of the target contaminant, works well compared to other techniques
Efficient	Removes target contaminant to a high degree; removes a higher percentage of the contaminant than other techniques
Cost effective or cost efficient	Removes more mass per unit of money spent ($/kg)
Simple to install and operate	Implementation does not involve significant effort
Dependable	System requires less maintenance and has lower downtime than other methods
Tested, accepted	Remedial method has been widely used and has shown good results; other methods may be more experimental, and thus have higher risk of failure

In the same way, economists refer to an analysis conducted by a problem holder or project proponent of their own direct costs and direct returns as a *financial analysis*. By definition, an economic analysis must also include external costs and benefits.

8.1.2 COMMUNICATION BETWEEN DISCIPLINES

The literature on the economics of groundwater protection and restoration has been divided into research published by environmental professionals and work published by economists. The two groups tend to publish in different journals and symposia, use different terminology, and focus on different aspects of the issue. The gulf between these groups can be narrowed in future through support for joint research. In addition, compartmentalization and specialization of economic studies in the realm of resource valuation, for example, may mean that economists are limiting their ability to help policy makers integrate information about the value of groundwater and the costs of impacts on the resource. Application of economic techniques to remedial decision making is considered by the authors to be necessary and valuable. Rapprochement of the various disciplines and subdisciplines involved in the issue will help to ensure that appropriate objectives for groundwater remediation are set and that the most cost-effective approaches and techniques are employed.

Here, the semantics are important. In applying a true economic analysis, the definitions used by economists need to be used and understood. In communicating between disciplines, a standard terminology is required. Ultimately, if arguments based on economic analysis are to be effectively presented to decision makers, they must be in a consistent language that everyone understands. Throughout this book, the economists' definitions are adhered to.

8.1.3 LITERATURE REVIEW — COSTS OF GROUNDWATER
REMEDIATION

The costs of actually implementing technical remedial solutions at specific sites where groundwater contamination exists are relatively well documented. The USEPA, for instance, provides a comprehensive guide on the costs of implementing various remediation techniques at sites across the U.S.[2] Here, the literature is almost exclusively found in the technical (scientific and engineering) realm; very little on this aspect of the problem is discussed in the economic literature. In addition, it is the authors' experience that many environmental consultancies around the world, and some of the major corporations involved in managing and remediating contaminated sites, have developed extensive databases on the costs of various remedial techniques for groundwater.

Not surprisingly, pump-and-treat (P&T) seems to be a favorite datum for comparisons with newer or more innovative remedial technologies. This is largely due to the significant number of P&T case histories available (mostly from the U.S.), many of which include cost (and sometimes effectiveness) information. Unfortunately, many of the P&T remedial programs undertaken in the U.S. over the last 20 years have not proved successful in terms of meeting the original remedial objectives. However, the consensus of current research into P&T is that, in very many cases, P&T was applied incorrectly or was being asked to achieve an objective to which it was not suited.[3] This illustrates a major weakness in much of the literature that compares the costs of various techniques — the methods being compared do not actually perform the same remedial functions or were not designed to achieve the same results. In assessing such studies, it is important to make the distinction between remedial approaches and remedial technologies.

Nyer and Rorech provide an overview of the elements of P&T systems designed to treat BTEX (benzene, toluene, ethylbenzene, and xylenes) contamination in groundwater.[4] Indicative costs of each component are provided, along with recommendations on cost-saving measures. Gatliff compared the cost of phytoremediation using selected species of trees to deal with shallow groundwater contamination by nitrates, pesticides, and heavy metals to traditional pump-and-treat techniques.[5] Sittler and Peacock considered different applications of air-sparging for achieving different groundwater remedial objectives, and so compared the costs and effectiveness of one technology applied in different ways.[6] O'Hannesin examined the costs of permeable reactive barrier systems for remediation of VOC's in groundwater.[7]

The intent here is not to provide a comprehensive review of actual recorded and reported costs for various remedial technologies. There is sufficient literature in existence already on this topic. Furthermore, costs of application can vary considerably from place to place and time to time, depending on the unique circumstance of each site, the jurisdiction under which the remediation takes place, and the appropriateness of the remedy to the problem (inappropriate application of a remedial technology will often skew its cost to the upper end of the scale, distorting the true picture). However, the key to stress is that cost information is widely available for remedial techniques, both in the literature and notably in the data banks of practitioners worldwide. For each situation, and for each cost–benefit analysis, costs of various alternatives need to be worked out based on site-specific conditions.

8.1.4 Private Costs of Remediation

Most of the available groundwater literature dealing with costs of remediation focuses on the problem holder's costs, known also as *private costs of remediation*. This is in part because private firms have developed considerable experience and knowledge of their costs and historically have had little impetus to focus on the wider issues. Interestingly, and despite this, very little literature actually exists on the private benefits of groundwater contamination and remediation. The subject of benefits is covered in more detail in Chapter 9.

Both the private (internal) and wider social (external) costs should be considered during remedial decision making and setting of remedial objectives. In situations in which the polluter has been identified as a private entity, the costs of implementing remediation will be borne wholly or substantially by that entity. However, society may also share some of the burden of cost of the remediation, should unmitigated effects to the wider environment occur as a direct result of the remediation. These are called the *external costs of remediation* and are discussed in the following sections.

The private costs of groundwater remediation typically may include any or all of the following, but in any case, should include all expenditure required to achieve the desired remedial objective:

- Site investigation and data collection costs, including the costs of performing nonintrusive surveys, drilling and installing monitoring wells, and sampling and analysis
- Data interpretation and analysis costs, including reporting, predictive fate and transport modeling, and risk assessment
- Decision-making costs, including economic analysis costs, negotiations with regulators, public meetings and information costs, and public relations costs
- Remedial design fees
- Legal fees
- Permitting fees
- Capital costs of the remediation system
- Operation and maintenance costs, including spare parts, power, labor, security, water and waste disposal, and taxes
- Disposal and waste management charges, such as costs for disposing of waste materials and by-products of the treatment system, tipping, and charges for disposal of recovered contaminants
- Remedial system modification costs and contingencies
- Validation costs, including sampling and reporting
- Closure costs
- Insurance

Costs of remediation should include capital and operation and maintenance costs discounted at the chosen discount rate, over the chosen planning horizon.

The main focus of this book is on remediation of groundwater that has already been impacted or that could be impacted by an existing spill event. However, the methods and concepts discussed here are equally applicable to the protection of groundwater from future contamination.

Groundwater protection involves preventing future groundwater contamination, either by direct measures or through implementation of policy. Groundwater protection costs incurred by the private sector could include a large range of activities, such as:

- Implementing environmental management programs
- Environmental training for employees
- Environmental monitoring plans
- Investment in plant and equipment designed to reduce the probability of leaks and spills that may result in groundwater contamination
- Development of spill-response plans
- Remediating soil contamination
- Removing, stabilizing, or isolating wastes
- Improving waste disposal practices
- Forgoing development to comply with groundwater source protection zone restrictions

Raucher comments that, in general, the value (or benefit) of avoiding a groundwater contamination incident will be at least as great as the expected cost of damage incurred should it occur.[8] The costs of groundwater remediation, in general, will tend to increase with the complexity and heterogeneity of the subsurface; the depth of groundwater; the mass, longevity, mobility, and toxicity of the contaminant; and the time until detection and action. Thus, groundwater contamination can be extremely expensive to remediate once it has occurred, and in some cases is irreversible. In contrast, many of the most effective spill prevention and groundwater protection measures are relatively inexpensive (training programs, environmental management programs, improved inventory, storage and handling practices). This confirms the widely held view that in many cases, prevention of groundwater contamination will be much more cost effective than remediation.

8.2 EXTERNAL COSTS OF REMEDIATION

8.2.1 OVERVIEW

As discussed in Section 8.1, the act of remediation can cause secondary effects that may result in an environmental impact in their own right, despite our best attempts at mitigation. These effects must be included in the overall economic assessment, if optimal remedial decisions are to be made. If the costs of dealing with these effects or the damages they cause are imposed on third parties (other than the problem holder) and are not compensated, they are termed *external costs of remediation*. Examples of some typical external costs of remediation include:

- Creating a new risk — In situations in which contaminants are removed from groundwater and introduced into another medium, a new risk that did not previously exist may result. For example, air-stripping of volatile organic compounds (VOCs) from groundwater, without the use of off-gas treatment, puts VOCs directly into the air, where they may affect the health of nearby residents. In some jurisdictions, this is allowed. Moving recovered contaminants to another location (such as a tip or landfill) could expose people along the transport route. Exposing remediation workers to risk (health and safety issues) is also an important potential cost of remediation — though as long as these risks are known to the workers, they could be assumed to be compensated by the wages (see Chapter 4).
- Contamination of another medium — Certain remedial approaches may involve redirecting contamination to another medium, such as soil, air, or surface water. Examples would include *in situ* volatilization processes, which drive volatile contaminants from water into unsaturated zone soils, discharge of pumped contaminated groundwater to the sewer system or to a wetland for treatment, or discharge of volatile compounds to the air.
- Contributing to air and greenhouse emissions — Any project that is energy intensive or produces inordinate levels of greenhouse and other air emissions through the remedial process itself may also be producing external costs associated with climate change and air pollution.
- Permanent elimination of water from the hydrologic cycle — If we assume that fresh water has some value, then a remedial process that removes it completely from the hydrologic cycle would produce a loss equivalent to the value of the volume of water processed. An example is deep-well disposal of contaminated groundwater, a common practice in many parts of the world.

8.2.2 Net Benefits and External Costs

If the prevention or remedial actions taken produce a secondary impact, this impact should be included in the analysis as an eternal cost of remediation. External costs of remediation are conceptually similar to negative benefits, in that they are damages that accrue over time, as impacts on stakeholders or resources. Thus, as long as they continue, they will accumulate. External costs of remediation can be expressed as X_{ARt}, depending on the nature of the effect produced by the remedial activity. It is assumed that these costs reflect the residual damages that occur as secondary impacts of the main remediation, after the application of available mitigation measures. Thus, the cost of implementing these mitigation measures (C_{ARmt}) is also included in the resulting revised NPV expression (Equation 8.1):

$$NPV_{ARt} = \sum_{0}^{T} \left[\frac{D_t - D_{ARt} - C_{ARt} - C_{ARmt} - X_{ARt}}{(1+r)^t} \right] \tag{8.1}$$

External costs of remediation can be divided into two categories:

- Planned or process-related external costs that cannot or will not be mitigated against (X_p)
- Unplanned or inadvertent or unforeseen external costs (X_{up}), as in Equation 8.2:

$$X_{ARt} = X_u + (p. X_{up}) \tag{8.2}$$

where p is the probability that the unplanned external cost will occur. Each of these is discussed below.

8.2.2.1 Planned External Costs

Many of the remedial solutions used today involve some degree of unmitigated secondary damage or liability to the environment or to another party. Depending on the regulations being enforced in a particular jurisdiction, these can vary from minor residual effects to major planned transfers of cost.

A good example is the common practice of landfilling wastes excavated from contaminated sites. Removal of concentrated zones of soil contamination by excavation is a common way to remove the source of ongoing groundwater contamination. Contaminated soil seen to be driving unacceptable risk at the site is excavated and transported to a secure landfill facility. Of course, the definition of *secure* varies, depending on which part of the world is being considered. Even in developed and well regulated jurisdictions, landfills have been known to leak and themselves become sources for subsurface contamination. Excavation and landfilling (or *dig-and-dump*, as it is often known) can cause external costs in the following ways.

- Potential damage to the receiving environment at the landfill — Unless the landfill site is 100% secure, there is the potential that the wastes may cause leachate or vapor releases, which may contribute to impacts on the surrounding environment, particularly groundwater. There is ample evidence in the literature to suggest that many of the landfills that have accepted or are accepting hazardous waste in developing countries are sources of impacts to groundwater. Recent studies in the U.K. indicate clearly a direct link between proximity to a landfill and a reduction in property value.[9] Although the overall impact of the remediation may have been positive, in that there has been an overall net reduction in the potential for the waste to harm human health or the environment, now that it is sequestered in a landfill, it remains as a potential source of long-term risk. The wastes have simply been transferred from one location to another, and now must be managed and contained over the long term. Valuation of this cost is difficult in practice. A highly conservative approach would be to include as external costs the cost of destroying or otherwise rendering the wastes inert through some form of *ex situ* treatment. The fees paid by the problem holder to the operator of the landfill represent the costs of mitigation against release of any portion of the waste into the environment,

TABLE 8.2
Valuation of Road Congestion

Reference	Impact Type	Change Being Valued/ Valuation Scenario	Valuation Method	Results
Newbery[10]	Congestion	HGV traffic marginal cost of congestion on different types of road based on economic value of time — United Kingdom	Mix	(U.K. £/HGV km) Motorway £0.006 ($0.0108) Urban central peak £0.8347 Urban central off-peak £0.6708 Noncentral peak £0.3639 Noncentral off-peak £0.2006 Small town peak £0.1581 Small town off-peak £0.0964 Other urban £0.0019 Rural main £0.0016 ($0.0029) Other trunk £0.0044 ($0.008) Other rural £0.0012 ($0.0022)

where it could cause damage. The external cost represents the residual impact, based on the probability of release of part of that waste, despite containment efforts.

- External costs of transporting waste to landfill using heavy goods vehicles (HGVs) — These can be estimated, based on the assumption that increased HGV traffic on the road network will result in a number of costs: increased congestion, impacts on health from emissions, noise impacts, and increased probability of accidents. The example that follows describes how the external costs of transport would affect the economic analysis of a remediation using dig-and-dump.

8.2.2.2 Example: External Costs of Transport to Landfill

In this example, the expected journey for road transport of excavated waste from a remediation project will make use mainly of trunk roads and highways. The distance from the site to the secure landfill is 675 km, and assuming the vehicles complete a return journey, the total journey length is 1350 km. It is expected that 10,000 tons of contaminated soil need to be excavated and removed to landfill. At an average of 10 tons per heavy goods vehicle (HGV), 1000 vehicle movements will be required to complete the work.

Table 8.2 through Table 8.5 provide information on studies investigating costs due to congestion, noise, health, and accidents. An external cost for each impact is

TABLE 8.3
External Costs of Road Transport — Health and Noise — 2002$USD/km

	HGV	Bus/Coach	Passenger Car
Health costs*	0.6309	0.4206	0.0421
Noise costs	0.0412	0.2754	0.0139

* Pollutants considered: PM10, SOx, NOx, VOCs, lead, benzene.

Source: From Maddison et al.[11] (UK£ = 1.8 US$.)

TABLE 8.4
Accidents Associated with Heavy Vehicle Traffic in the U.K. 2002US$/km

Total fatalities due to HGVs in 2001 (A)	588
Total serious injuries due to HGVs in 2001 (B)	2,910
Total vehicle kilometers by HGVs in 2001 (billions) (C)	29.2
The average risk of fatality per vehicle km for HGVs (D)	0.20×10^{-7}
The average risk of serious injury per vehicle km for HGVs (E)	1.00×10^{-7}
The cost of serious injury measured as the WTP to avoid injury (F)	$ 177,220

Source: From DfT.[12]

TABLE 8.5
Value of a Statistical Life — 2002US$

The cost of fatality measured as the value of statistical life (G)	4,206,438
Total cost of fatality per vehicle km for HGVs (D $\times$ G)	$0.08
Total cost of serious injury per vehicle km for HGVs (E $\times$ F)	$0.02

calculated and then the three external costs are aggregated to provide a total unit cost for heavy vehicle movements. Note that the costs in this example involve transport of the soil alone and not the external costs associated with landfilling.

8.2.2.2.1 Road Congestion

The remedial program will result in 1000 additional long road journeys that otherwise would not have been undertaken. This adds to the congestion of roads along the planned route and has a measurable economic impact. Table 8.2 shows relevant information on valuation of road congestion.

Considering the mix of road types used in the journey provided in Table 8.2 (movements: motorway [28%], trunk [57%], rural dual carriageway [13%]. and other rural [2%]), a unit cost for congestion per vehicle per km can be calculated:

Unit cost of congestion per HGV km:

$$= (\$0.0108 \times 0.28) + (\$0.008 \times 0.57) + (\$0.0029 \times 0.13) + (\$0.0022 \times 0.02)$$

$$= \$0.003 + \$0.0045 + \$0.0004 + \$0.000044$$

$$= \$0.008 \text{ per vehicle-km } (\$2001)$$

$$= \$0.009 \text{ per vehicle-km } (\$2002)$$

8.2.2.2.2   Health and Noise

The planned truck movements will generate additional noise and additional health impacts, primarily as a result of air emissions from exhaust and dust. These impacts will be felt by people living along the planned route. These people have no involvement in the remediation program and will not benefit directly from the remediation. Thus, part of the cost of remediation is imposed on them and must be accounted for in a complete economic analysis. Table 8.3 details results of recent valuation studies of these impacts.

Applying the values in Table 8.3 to the example, unit external costs of transport from noise and health impacts resulting from the vehicle movements for transport of excavated soil to landfill would be estimated as:

Unit cost of health and noise per HGV km:

$$= \$0.6309 + \$0.0412$$

$$= \$0.672 \text{ per vehicle-km } (\$2002)$$

8.2.2.2.3   Accidents and Fatalities

The additional truck movements will bring a statistical probability (albeit very small) that an accident involving one of the vehicles will occur during the course of the remediation. Statistics are available that allow the cost of this possibility to be estimated, as shown in Table 8.4.

Using the data in Table 8.4 in conjunction with an estimate for the value of a statistical life presented in Table 8.5, an estimate for the unit cost of a fatality and serious injury per vehicle-km is calculated.

Unit costs of accidents per vehicle-km:

$$= \$0.08 + \$0.02$$

$$= \$ 0.1 \text{ per vehicle-km } (\$2002)$$

8.2.2.2.4   Total External Cost of Transport

Using all of the preceding information, the total external costs for transporting the material to the landfill using heavy goods vehicles can be calculated as the sum of congestion, health, noise, and accident-related external costs:

Total unit external costs of transport per vehicle-km:

$$= \text{costs of congestion} + \text{health} + \text{noise} + \text{accidents}$$

$$= \$0.009 + \$0.672 + \$0.08 + \$0.02$$

$$= \$0.78 \text{ per vehicle-km } (\$2002)$$

Using the unit cost of $0.78/vehicle-km, the total external cost of transporting the excavated contaminated soil from the site to the landfill can be estimated. For 1000 vehicle movements, each of 1350 km round trip, a total cost of $1.053 million is calculated. This sum represents the costs borne by others as a result of the additional vehicle traffic that resulted from the decision to use the dig-and-dump method to remediate this site.

The relevance of this impact can be seen by considering typical private remediation costs for excavation and landfilling of 10,000 tons of contaminated soil. A typical remediation program of this size would cost in the order of $2 to 5 million, depending on location, contaminant type, tipping fees, and the complexity of the dig. In this example, the expected private or internal cost for remediation using dig-and-dump was expected to be approximately $3.2 million. Adding $1.05 million to reflect the real cost of the remedy represents a 33% increase. Note also that if clean fill has to be imported to site to fill in the excavation, additional vehicle movements will be required, further boosting the external cost of transport. Furthermore, the other possible external costs of landfilling have not yet been added.

Examples of other types of planned external costs of groundwater remediation are listed in Table 8.6. In general, planned external costs are increasingly being mitigated against. In many jurisdictions, specific regulatory measures are being put in place to ensure that remediation methods that deliberately shift costs from the problem holder to society are reduced or eliminated. A recent example is the Euro-

TABLE 8.6
Examples of Planned External Costs of Remediation

Activity	Secondary Effect	Comments
Air-stripping of volatile compounds from groundwater without off-gas treatment	Release of volatile compounds to atmosphere	Still occurs in many jurisdictions, can be mitigated against
Thermal treatment of contaminated soils	Release of CO_2 and other gases to atmosphere	Greenhouse gas emissions
Permanent sequestering of contaminated groundwater	Permanent loss of injected groundwater as a resource	Widely used for difficult and recalcitrant contaminants
Excavation of concentrated source of contamination to protect underlying groundwater results in habitat destruction	Habitat in excavated area destroyed	Mitigation banking approaches can be used to offset

pean Union's Landfill Directive, which puts stringent new limits on landfilling of hazardous wastes excavated from remediation sites.

Accounting for unplanned or unforeseen costs of remediation is, of course, problematic — we may not know that they are going to happen, or we may have discounted them as only a remote possibility.

8.2.2.3 Unplanned External Costs

Sometimes, despite best planning and care, remediation activities result in creation of a secondary impact to the environment or to other stakeholders. If the impact is an unplanned or unforeseen result of remediation, for which mitigation measures have not been provided or have not been successful in countering, then the value of this damage is included as an external cost of remediation. Table 8.7 provides a list of examples of unplanned external costs.

Accounting for unplanned external costs within an economic evaluation of remedial alternatives is not straightforward. For any given remedial approach being considered, the possibility that its implementation may cause additional external

TABLE 8.7
Examples of Unplanned External Costs of Remediation

Activity	Secondary Effect	Example
Remediation causes LNAPL to revert to DNAPL, due to preferential removal of lighter compounds	NAPL sinks, contaminating a new volume of aquifer, worsening dissolved phase problem	SVE (soil vapor extraction) preferentially removes volatile aromatics from an LNAPL containing less volatile dense compounds
Bioremediation results in creation of daughter products which are more toxic than parent	Toxicity to receptors increases	TCE (trichloroethene) degrades to VC (vinyl chloride), and VC persists in aquifer
Remediation inadvertently increases mobility of contaminant within the aquifer, through alteration of physiochemical properties	Impact on receptors worsens, due to further spreading of plume, increased mass flux, or more rapid breakthrough	Surfactant flush greatly increases dissolution of NAPL, containment insufficient
Remediation inadvertently increases mobility of contaminant within the aquifer, through alteration of properties of the aquifer itself	Impact on receptors worsens, due to further spreading of plume, increased mass flux, or more rapid breakthrough	In situ fracturing of aquifer to enhance NAPL recovery inadvertently allows increased NAPL mobility toward receptors
Remediation compromises adjacent confining layers or geological features	Contaminant is introduced into a hitherto uncontaminated geologic unit	Pumping wells completed across a confining layer, cross-connecting two groundwater-bearing zones

damages must be carefully evaluated. In most situations, experienced remediation engineers and specialists should be able to identify possible secondary damages. In all cases, mitigation measures should be put in place to deal with these possibilities. Whatever probability remains of that damage occurring after the mitigation measures are put in place (and the cost of their implementation added to the overall cost of remediation) should be included as part of the external costs (see Equation 8.2). Assigning a probability to an eventuality that is being mitigated against is a matter of the remediation team's professional judgment and should be based on experience and on knowledge of the limitations of remedial technologies and the mitigation measures themselves.

8.2.2.4 Example: Accidental Piercing of Basal Containment Layer

A site is contaminated by DNAPLs (dense nonaqueous phase liquids) that have migrated from a number of subsurface storage tanks into a shallow, unconsolidated gravel aquifer. Over time, DNAPL has accumulated atop a thin, low-permeability clay unit that overlies and hydraulically isolates an underlying fractured carbonate aquifer used extensively for local water supply. Several public water supply wells pump from the bedrock aquifer. The thickness of the clay unit directly beneath the DNAPL accumulation is not known; however, angled drilling has confirmed that, to date, DNAPL has not penetrated into the underlying bedrock. Drilling at other locations nearby suggests that the clay thickness varies from about 0.5 to 2 m. Some drilling logs from other sites nearby have even failed to notice the clay. Indeed, presence of dissolved-phase DNAPL components within the bedrock further down-gradient of the site suggest that windows through the clay may exist, allowing dissolved-phase contaminants emanating from the DNAPL and migrating with the gravel aquifer to find their way into the bedrock.

One remediation approach being considered includes removal of the bulk of the DNAPL from the gravel aquifer, to prevent further dissolved-phase migration within the gravel, reduce risk to public water supplies using the bedrock aquifer, and make the site suitable for redevelopment. Costs for this remedial approach were developed assuming excavation and removal of underground tanks and DNAPL-contaminated gravels associated with these sources, followed by on-site soil washing. Excavation of all of the DNAPL-contaminated material would require piling into the clay unit, to prevent the inflow of groundwater, and allow removal of the contaminated gravels. After washing, clean gravels would be replaced in the excavation and contaminated fines landfilled. Table 8.8 shows estimated costs of about $3 million for this option.

For a complete economic analysis of this option, however, the external costs must be considered. As discussed earlier, landfilling the fines will create a planned external cost, based on the residual liability presented by this material in its new location, and the external costs of transportation. In this case, the landfill was close to the site, and the volume involved was small, making its impact on overall costs negligible. However, there is real risk that piling into the clay unit may open up pathways for DNAPL migration into the bedrock aquifer below. If this were to occur, a significant dissolved-phase plume would develop in the fractured aquifer, which

TABLE 8.8
Cost Estimate — Excavation and Treatment of Contaminated Gravels

Item	Basis	Cost ($M)
Piling	(8 m deep, × 340 linear m) interlocking	0.77
Excavation	Excavate gravels, replace clean material	0.10
Water treatment	Dewatering systems, air-stripper, carbon filter, discharge to sewer	0.18
Gravel washing	Wash 10,000 m³ of gravels	0.93
Landfilling fines	Landfill 2,000 m³ of fines	0.36
Professional fees	Design and supervision, validation, reporting	0.45
Total Cost	With contingency (7%)	3.00 approx.

modeling results show would impact the nearest public water supply well within 100 days, at concentrations that would knock the well out of production. What is more, removal of the DNAPL from the fractured bedrock would be extremely difficult, if not impossible.

To allow an economic comparison of this particular remedial option with other options, a full accounting for the costs of remediation must be made. In this case, the possibility that an unplanned external cost could occur is a critical issue that cannot be ignored. One could argue that a judgment could be made that the risk to the bedrock aquifer was simply too great, and that the option should be discarded altogether. However, the same could be said of any number of remediation measures. Instinctively, remediation engineers make judgments on residual risks of remediation, but generally they do so without explicitly considering the economic implications of those risks. In just as many cases, remediation engineers make judgments that the risks associated with a particular option are acceptable. An explicit and complete analysis of the costs and benefits of each viable remedial option allows those risks to be put into perspective, eliminating to some extent the reliance on judgment and providing a clear record of the decision process and quantifiable justification for decisions.

In the example, discussions with piling contractors indicated that a small risk of penetration of the clay layer was possible, either through overpiling, or through pushing large cobbles ahead of the piles, which could penetrate the clay in advance of the piles. This risk was exacerbated by the fact that the thickness of the clay in the zone of piling was not known (and could not be determined due to the risk of drilling through the clay and introducing DNAPL into the bedrock that way). The probability of piercing the clay during piling was conservatively set at 25% for the analysis.

If NAPL were to enter the underlying aquifer, probabilistic groundwater modeling indicates that expected concentrations of key dissolved-phase contaminants in the public supply well would reach up to 1000 times the drinking water standard within a few weeks. This would necessitate an emergency response plan, to replace the lost production from the well in the short term, and a longer-term treatment requirement, including installation of new treatment systems and operation of those systems over the length of the planning horizon (20 years) and beyond.

The external cost of DNAPL introduction into the bedrock aquifer can be valued as the cost of protecting the users of the public water supply (PWS) well that would be affected by the contamination, plus the economic value of that part of the aquifer rendered unusable. A summary of the external costs calculated is provided in Table 8.9. Assuming an immediate response involving a replacement supply of the current production of 4.16 million m^3/yr at market rates, over a 3-month period, a notional cost for emergency response of $1.6 M can be estimated. Assuming a unit treatment cost of $0.18/$m^3$ over the remainder of the planning horizon, discounted at 3.5%, an additional PV cost of $10.6 M is estimated. For this example, the value of the lost aquifer potential was not estimated, because the PWS in question is the major and only licensed user of the aquifer in this area and hence will be treated rather than shut. Total present value (PV) cost of NAPL contamination of the bedrock aquifer would be estimated to be in the order of $12.2 M (Table 8.9).

Using the probability factor of 0.25 and Equation 8.2, an external cost of remediation (X_{up}) of $12.2 \times 0.25 = $3.06 million is estimated. From Table 8.8, the estimated cost for the excavation option was approximately $3.0 million. The possibility of unplanned external costs, in this example, raises the total cost to $6.08 million, an increase of 100%. Even dropping the probability factor to a less conservative 0.1 (10%), results in an overall increase in remedial cost for this option of about 40%.

8.2.3 POLLUTION LIABILITY INSURANCE

In some situations in which the possibility of external costs is real, proponents may choose to avail themselves of some of the insurance instruments currently available in the marketplace. Bespoke project-specific pollution liability insurance policies are available from several of the major underwriters. Premiums and deductible amounts for these policies will depend on the nature of the risk, the contaminants involved, the details of the site and the potential receptors, and the reputation and technical competence of the consultants and contractors involved in the planned remediation. Each case is evaluated on its own merits.

However, from a purely economic perspective, the availability of an insurance policy does not in itself reduce the potential of damage to the environment. The damage still occurs, but financing mitigation or further clean-up becomes an easier task for the problem holder who benefits from transferring the risk onto the insurer

TABLE 8.9
Possible External Costs of Inadvertent Contamination of Bedrock Aquifer

Item	Basis	PV Cost ($M) @ 3.5%
Emergency response	3 month equivalent water delivery	1.58
Long-term treatment at PWS	20 years; 4.16 Mm³/yr at $0.18/m³	10.6
Total		12.2

should a problem occur. As a contributing member of society, the insurance company itself becomes a stakeholder. Recall from Chapter 4 and Chapter 5 that, like taxes and subsidies, insurance fees are also transfer fees — from one party in the society to the other with no net change in the society's welfare — and hence are not included in an economic analysis (though are included in a financial analysis as items of cost to the problem holder).

REFERENCES

1. James, B.R., Huff, D.D., Trabalka, J.R., Ketelle, R.H., and Rightmire, C.T., Allocation of environmental remediation funds using economic risk-cost-benefit analysis: a case study, *GWMR*, 16, 4, 95, 1996.
2. USEPA, *EPA Guidebook: Remedial Action at Waste Disposal Sites*, USEPA, EPA/625/6-85/006, Washington, DC, 1995.
3. Hoffman, F., Groundwater remediation using "smart pump-and-treat", *Groundwater*, 31, 11, 98, 1993.
4. Nyer, E.K., and Rorech, G., Treatment system operation and maintenance: critical factors in an economic analysis, *GWMR*, 12, 4, 97, 1992.
5. Gatliff, E.G., Vegetative remediation process offers advantages over traditional pump-and-treat technologies, *Remediation*, 4, 3, 343, 1984.
6. Sittler, S.P. and Peacock, M.P., Innovative air sparging modifications to remediate organic compounds in non-optimum geologic formations, *Proc. 11th Nat. Outdoor Action Conf.*, Las Vegas, 259, 1997.
7. O'Hannesin, S.F., Long-term cost advantages of permeable reactive barrier technology for the remediation of VOC contaminated groundwater. *Proc NGWA Conf. on Remediation: Site closure and the Total Cost of Cleanup*. Nov, 2003. New Orleans, 303.
8. Raucher, R.L., A conceptual framework for measuring the benefits of groundwater protection, *Water Resour. Res.*, 19, 2, 320, 1983.
9. DEFRA (Department for Environment, Food and Rural Affairs), *A Study to Estimate the Disamenity Costs of Landfill in Great Britain,* final report by Cambridge Econometrics with eftec and WRc, London, 2003.
10. Newberry, D., Economic principles on pricing roads, *Oxford Rev. Econ. Policy*, 6, 2, 1992.
11. Maddison, D., Pearce, D.W., Johansson, O., Calthrop, E., Litman, T., and Verhoef, E., *The True Costs of Road Transport*, Earthscan, London, 1996.
12. U.K. Department for Transport (DfT), *Transport Statistics Great Britain*, Department for Transport Web site (www.dft.gov.uk), 2002.

9 Remedial Benefits

As also defined in Chapter 4 and Chapter 5 of this book, Abdalla argues that, from a public decision-making standpoint, the benefits of groundwater protection can be viewed as damage avoided.[1] Major damage categories can be categorized after the nomenclature of Spofford.[2]

- Human health effects, occurring due to exposure to contaminated groundwater
- Potential "stigma" damage to the value of properties seen to lie on a contaminated site or affected in some way by subsurface contamination
- Ecological damage and loss of recreational use, stemming from groundwater's role as a contributor to surface water flows
- Reduction or loss of nonuse values, through impact on option or bequest value

The preceding list is not exhaustive, and other types of benefits may accrue, depending on the characteristics of the groundwater, contamination, and the sites affected. Whatever their source, the benefits of protection and remediation are grouped into two categories:

- Private or internal benefits
- Public or external benefits

The terminology *internal* and *external* is borrowed from the economic literature and is also discussed in Chapter 4 and Chapter 5.

9.1 PRIVATE (INTERNAL) BENEFITS

Most of the literature reviewed in Chapter 4 deals with the costs and benefits of remediation that affect the wider society (beyond the problem holder). This is in part because private firms have developed considerable experience and knowledge of their costs and benefits in this area, so researchers have tended to focus on the wider issues. Despite this, little literature actually exists on the private benefits of groundwater contamination and remediation.

However, both the private (internal) and larger public (external) costs and benefits should be considered during remedial decision making and setting of remedial objectives. The private benefits of remediation could include these listed in Table 9.1.

These benefits can legitimately be included in a private (internal or financial) analysis of a remediation decision. However, as Chapter 5 shows, not all of the private benefits are net increases in the wealth, well-being, or welfare of society but simply transfers from one party in the society to another. For examples, fines avoided

---

TABLE 9.1
Private Benefits of Remediation

Direct Benefits of Remediation	Costs Avoided if Remediation Takes Place
• Sale value of cleaned-up site • Costs savings through access to clean groundwater (if aquifer used by the problem holder) • Improvements to "green image" of company/organization	• Avoided risk of litigation (and hence costs involved in litigation) • Avoided financial liability • Avoided fines • Avoided public relations damage • Avoided loss of shareholder value • Avoided control orders or shut-downs • Avoided exposure of on-site personnel to pollutants

---

are benefits of remediation from the point of view of the problem holder but not from the point of view of the whole society, because saved fines for the problem holder are lost revenue for the competent authority imposing those fines.

It is clear from the preceding list that some of the private benefits are relatively easy to quantify. For example, potential income from the sale of cleaned land can be quantified by analyzing the property market in the area concerned. Property agents maintain a detailed and thorough listing of property values and selling prices. This allows ready monetization of this private benefit of remediation. Other types of benefits, such as potential public relations damage, however, could be much more difficult to quantify given the uncertainty about the reactions of customers and shareholders.

Being one of the most robust, relatively straightforward, and already familiar ways of examining the economic impact of site contamination, property values are worth considering in more detail. In simplified terms, the benefit of remediation is the difference between the value of the property before remediation and its value after remediation. In many places, the value of property drives efforts to remediate and then sell the land. The intimate relationship between contaminated sites and the groundwater that lies below, as discussed previously, means that property value can be an important part of remedial benefits. Cleaning up property and sites that act as long-term sources of groundwater contamination leads to direct benefits associated with groundwater itself, but also to benefits associated with the property and surrounding properties.

First, there is the value of the site itself. A contaminated property will almost always sell at a discount of the price that could have been fetched if the site were fit for purpose. In many cases, one of the key benefits of remediation to the owner of a contaminated site is the increase in property value achieved. If the increase in value is greater than the costs of remediation, a net profit is realized. This can be a powerful impetus to clean up sites, and this is why, throughout the developed economies, a growing number of brownfield developers are seeking to capitalize on the often considerable price margin between "dirty" and "clean" sites. *Brownfield projects* typically involve purchasing a contaminated property at a substantial discount, remediating the site, and then selling the property for more than the sum of the original purchase price and the remedial cost. Proponents of these projects take on the liability for the contamination under the assumption that they can remediate the site, eliminate or manage the liability, and produce a site that can be sold at or near the full market value.

Different organizations and individuals have different levels of risk tolerance, so their willingness to pay for liability reduction can be markedly different. Firms or groups that understand the technical, legal, and financial complexities of the contaminated land business are more likely to be tolerant of environmental risk and liability, and are better placed to execute a brownfield project profitably. Because market prices are often driven at least partially by perceptions of risk and these perceptions can vary considerably, regulators have a key role to play in brownfield transactions. Regulatory approval of remedial designs and results can be instrumental in creating comfort in the market that remediation has been successful, thus unlocking site value.

9.2 PUBLIC (EXTERNAL) BENEFITS

Public or external economic benefits of groundwater remediation arise from the avoided damage to the environment and to human health. The environmental damage could affect not only the flora and fauna and the health of the ecosystem but also individuals' consumptive or nonconsumptive (e.g., for recreation) uses of the affected resources. In addition, individuals may benefit from knowing that, due to remediation, they themselves or subsequent generations could use the aquifer in the future. These different ways in which remediation benefits accrue to the society other than the problem holder correspond to the components of the total economic value (i.e., use and nonuse values) (see Chapter 4).

These benefits are termed *external* because they are not included in the actual market transactions for land or water and in the financial analyses undertaken by problem holders.

The rest of the society that benefits from remediation could be individuals, households, or businesses (e.g., farmers who use the impacted well water for irrigation). The mechanism through which benefits accrue applies equally to these different parties of the affected population. Note that if damage (or cost) imposed on a particular user of the aquifer is compensated by the problem holder, this cost ceases to be external and becomes a financial (internal) cost to the problem holder. This is because externality requires the impacts (positive or negative) to be uncompensated. In that case, remediation would save the problem holder the compensation amount it has been paying and makes no difference to the welfare of the affected party receiving that compensation.

Chapter 4 shows the ways in which the external cost of damage (or external benefit of remediation) can be quantified in monetary terms and briefly reviews some examples from the current literature. However, as the chapter indicates, there are a number of obstacles to full valuation of the benefits of groundwater remediation and protection. These obstacles include:

- Difficulty and expense in monetizing site-specific benefits — The literature review revealed that most researchers feel that the economics of groundwater remediation is site specific. Clearly, then, valuation of all of the benefits of a project will involve considerable effort and no small amount of site-specific research. An economic data-worth analysis is probably worthwhile in many situations to determine the level of data required for appropriate decision making. It is considered unlikely, however, that sufficient time or resources will be available in most cases to develop fully monetized values for all categories. Smaller sites, with less serious contamination problems, will likely not warrant any site-specific valuation studies, and the external benefits of remediation may only be partially monetized. For larger, more heavily contaminated sites involving significant current and potential future risks, a more complete economic assessment may be worthwhile. This tiered approach of adjusting the level of analysis to suit the situation is favored as a practical way to include economic factors in the decision making without making the analysis overly onerous.
- Inherent limitations of economic valuation — Expressing all costs and benefits in monetary terms is not possible for several reasons:
  - Scientific information that is used to identify impacts may not exist.
  - This information may exist but may not be in units that are conducive to economic valuation.
  - Economic research may not have been done.
- Some of the most difficult-to-measure values may be among the most significant — Little research has been conducted in the nonuse value of groundwater. Despite this, there seems to be agreement among the work reviewed that the literature does indicate a strong WTP for nonuse benefits of groundwater. This fact alone may be sufficient to influence groundwater protection or remediation decisions. Some researchers feel that, if properly measured, nonuse values could be considerable. In practical terms, these views suggest that in many situations (in which only easily measured values are considered in a cost–benefit analysis), the benefits of remediating or protecting groundwater could be considerably underestimated. The significance of this underestimation depends mainly on the availability of substitutes for the affected aquifer, not only for human consumption but also for its role in the surrounding ecosystem — the fewer substitutes there are, the more the nonuse value is likely to be. Therefore, if the current trends of increasing pressures (both from increasing demand and pollution loads) continue, in time the nonuse values of groundwater are likely to increase.

Despite these obstacles, which can hinder a full quantification of external benefits of remediation, the literature offers information and values that are of relevance at least to illustrate methodologies. For example, the preceding section discussed increased property value due to remediation as a measure of private benefits. This is the case for the property or properties owned by the problem holder. However, when a contaminated site is remediated, the site owner is not the only one to benefit. Through the removal of what might have been an eyesore or a potentially hazardous condition, the whole neighborhood benefits.

Several recent economic studies have robustly shown that people and businesses perceive a real economic benefit when a neighboring waste site or polluted site is remediated. This is due to the removal of blight or disamenity from the properties in the vicinity of the remediated site. This effect is intuitive — people would rather live in an area without contamination and waste, if they had the choice. Recent research by Gawande and Jenkins-Smith found substantial negative effects on property values in areas subjected to transshipment of radioactive wastes in the United States.[3] Payne et al., found a 35% difference between average selling price of homes located within a 2-mile radius of a low-level radioactive waste site in the U.S., to those outside a 2-mile radius.[4] Hirshfeld et al., found decreases in property values between 12 and 25%, depending on distance from a hazardous waste landfill in the U.S.[5]

A recent study for Defra considered the disamenity costs of landfill in Great Britain.[6] The study considered 11,300 landfill sites and over half a million residential property transactions within 2 miles of those landfills, over the period of 1991 to 2000, inclusive. Residential property prices were found to be negatively affected within 2 miles of landfills. Across Great Britain, property values were found to suffer a 7% reduction within 0.25 miles, decreasing with distance to a 1% reduction between 0.5 and 1 mile from a landfill, and 0.7% between 1 and 2 miles from a landfill. In Scotland, however, impacts on property values were greater, decreasing 41% within 0.25 miles, 3% between 0.5 and 1 mile, and 2.67% between 1 and 2 miles. The term *disamenity* is generally used to define a number of impacts, such as noise, odor, litter, pests, and so on. However, the effect on property prices could also include people's perception of the risk to human health and the environment from shipment and disposal of waste or other sources of pollution. It is not possible, based on the currently available evidence, to differentiate this risk factor, and such differentiation is not strictly necessary, considering that remediation generally addresses both the actual impacts and the risk perceived (rightly or wrongly) by affected individuals as caused by the polluting activity.

Therefore, removal of the disamenity by remediation will cause average property prices in the affected area to rise. This increase, multiplied by the number of properties, can be used as a direct market valuation of disamenity or blight reduction and estimates the economic benefit that accrues to those stakeholders involved. This benefit will be greatest in dense urban areas with many neighbors and higher property values.

9.3 EXAMPLE — PROPERTY VALUE INCREASE AS A RESULT OF REMEDIATION

A site is contaminated with a variety of contaminants from a former manufactured gas plant (MGP) facility, including DNAPLs in shallow sediments, which are the source of an ongoing dissolved-phase plume in groundwater. The site is located in the heart of a busy town, in a prime commercial district bordering on an area of high-value housing.

Cleaning up the site and making it fit for resale will also go a long way toward achieving the goal of protecting local groundwater supplies being impacted by DNAPL. Two significant benefits of remediating the site will be the increased value of the property itself (which could be made suitable for sale and reuse) and increased value of the surrounding properties through removal of disamenity.

The full value of the site, remediated to a standard suitable for commercial redevelopment, was estimated at approximately $12.6 million. This estimate was based on market analysis completed by a local realtor, based on previous sales of similar properties in the current market. Techniques that use data from actual markets such as this are simple, quick, and robust (if incomplete). In this case, the total sale proceeds would be counted as private benefits to the problem holder (or site owner, assuming that the two are the same party). In other words, the site has no net value in its current state. Only through remediation can the value be unlocked. It is assumed that this value will be obtained only if the site is made fit for redevelopment through removal of substantial contamination and residual liability from the site. So, for example, pathway management objectives will not result in the accrual of this benefit. More aggressive source removal approaches will result in a higher proportion of this overall site value being realized.

Benefits of remediation to off-site residents are estimated by the increase in property value surrounding the site realized by remediation. Put another way, remediation will eliminate some or all of the negative effects on neighboring property values. This is an *external benefit* of remediation. For valuation purposes, the perception of damage to the surrounding people and lands and to public health (which is not actually affected in this case) can be assumed to be reflected by a blight effect on property values in the neighboring community.

On this basis, blight on property value is estimated conservatively by assuming that all of the values of all of the 200 residential properties (identified by aerial mapping), 100 apartment properties, and 100 commercial properties within a 500 m radius of the site are reduced by 5% because of the perception of damage and risk resulting from contamination at the site. Realtor databases were again used to assign average values to the different categories of neighboring properties.

Table 9.2 summarizes the external and internal benefits resulting from property value increase. Site remediation thus results in a present value benefit of $19.575 million, of which just under $7 million (just over a third of the total) is external, realized by the owners of properties surrounding the site.

9.4 TIME

The inherently dynamic nature of groundwater contamination has been discussed previously. It seems clear that the choice of planning horizon will have a significant impact on cost–benefit calculations for groundwater remediation. Over time, growth of a plume would tend to affect a greater number of receptors, resulting in increased

TABLE 9.2
Private and Public Benefits of an Example Site as Measured by Property Value

Benefit	Basis	Value ($M)
Site value increase (private or internal)	Sale of site for unrestricted commercial redevelopment	12.60
Blight removal — Residential properties within 500 m (public or external)	200 properties, average value $360,000; 5% blight factor removed	3.60
Blight removal — Apartments within 500 m (public or external)	100 apartment properties; average value $225,000, 5% blight factor removed	1.125
Blight removal — Commercial properties (public or external)	100 commercial properties; average value $450,000; 5% blight factor removed	2.25
Total benefit (in present value terms)		19.575

damage costs. Longer planning horizons mean benefits and costs are accrued over a greater number of years, resulting in greater total benefits and costs, all other factors remaining equal. Longer planning horizons may be required in situations involving deep subsurface sources of contamination or complex hydrogeological conditions. Irreversible contamination may simply have to be managed in perpetuity. In these situations, long-term operation, maintenance, and monitoring costs may be considerable, and the life expectancy of remediation system capital must also be considered (for example, the typical electric submersible pump deployed in a pump-and-treat application will need to be replaced every five to eight years).

This is why all benifits and costs should be entered into CBA in the year that they occur throughout the remediation project life time. Future values are then discounted to estimate present values using constant or declining discount rates as advised by official guidance (e.g., HM Treasury and USEPA).[7,8]

9.5 EDUCATION

Integrating of economic considerations and techniques into remedial decision making for groundwater problems is clearly a relatively new development, even in the United States, where there has been a long history of concerted action on groundwater contamination. In any new endeavor, success is predicated to some degree on the participants' levels of understanding and knowledge. If guidance on cost–benefit analysis for groundwater remediation is to be put in place and used effectively, those who are to use the methods and the stakeholder groups who will be asked to act upon the conclusions of the analysis will need a basic understanding of the main issues discussed in this book. In fact, basic knowledge of and training in economic methods are requirements for the EPA staff before they can implement the economic analysis guidance, as stated by the USEPA.[8]

REFERENCES

1. Abdalla, C.W., Groundwater values from avoidance cost studies: implications for policy and future research, *Amer. J. Agr. Econ.*, 76, 1062, 1994.
2. Spofford, W.A., Krupnick, A.J., and Wood, E.F., Uncertainties in estimates of the costs and benefits of groundwater remediation: results of a cost–benefit analysis, *Resources for the Future*, Discussion Paper QE89-15, Washington, DC, 1989.
3. Gawande, K. and Jenkins-Smith, H., Nuclear waste transport and residential property values: estimating effects of perceived risks, *J. Env. Econ. Manage.*, 42, 207, 2001.
4. Payne, B., Olshansky, S., and Segel, T., The effects on property values of proximity to a site contaminated with radioactive waste, *Nat. Resources J.*, 27, 579, 1987.
5. Hirshfeld, S., Veslind, P.A., and Pas, E.I., Assessing the true cost of landfills, *Water Manage. Res.*, 10, 471, 1992.
6. Defra (Department for Environment, Food and Rural Affairs), *A Study to Estimate the Disamenity Costs of Landfill in Great Britain*, final report by Cambridge Econometrics with eftec and WRc, London, 2003.
7. HM Treasury, *The Green Book: Appraisal and Evaluation in Central Government*, TSO, London, 2003.
8. USEPA, *Guidelines for Preparing Economic Analyses*, Office of the Administrator, USEPA, Washington, DC, 2000.

Part III

Remedial Decision Making

10 Context

This part of the book presents a step-by-step framework for making decisions on remedial objectives, approaches and technologies, taking into account the likely costs and benefits of each option. Chapter 10 provides the context for the framework, and provides an overview of how the framework fits together. Chapter 11 and Chapter 12 describe each step of the framework. Examples of its application are provided in Part IV.

10.1 LEVELS OF DECISION MAKING

The economics of groundwater remediation can be considered at four main levels:

- Policy level (government), which, for the purposes of this discussion, is considered set and is not considered in detail
- Remedial objective level
- Remedial approach level
- Technology selection level

The *remedial approach* is the link between the objective and technology levels, and must be considered at both of these levels.

In most cases, policy is considered to be a constant over the life of a given project. However, as discussed in Part II, regulations and laws do change over time. Increasingly, cost–benefit analysis is being used as part of regulatory impact assessments (RIA), in which new (or, in some cases, existing) government policies or regulations are evaluated in terms of the costs they will incur for implementation and the benefits they will produce *for society as a whole*. The aim of an RIA is to prevent new regulations from being enacted that will cost more than the benefits they deliver. Typically, RIA also considers who bears what proportion of the overall cost of implementation and which stakeholders reap the benefits.

10.1.1 POLICY OBJECTIVE LEVEL

Policy objectives are set by governments and are not the subject of this chapter. However, decisions on what to remediate, what to protect, and what to sacrifice must be generally guided by the policy of the day. Policy could include maximization of human welfare, for instance. In many jurisdictions, including the United Kingdom, Europe, Canada, and the United States, stated national environmental policy is based on the protection of human health. Standards, procedures, and guidance produced by the relevant authorities are designed to protect people from health impacts resulting from exposure to environmental contaminants.

10.1.2 REMEDIAL OBJECTIVE LEVEL

Setting the remedial objective (or risk management objective) for a given contamination problem should explicitly incorporate the standard risk assessment. Only a limited number of remedial (or risk management) objectives are available:

- Receptor is protected.
- Impacts to the receptor are reduced or eliminated.
- Contamination is removed.
- Contamination is reduced to a set, predetermined regulatory level. (This last objective is not formally risk-based.)

The remedial objective is the level at which the benefits of remediation are most readily and fundamentally determined. If a valuable receptor is protected, a benefit to society accrues as avoided damage. Not protecting the receptor results in damage. In this framework, benefits are clearly tied to the fundamental objective and the basic approach used to achieve it. For a groundwater contamination problem, for instance, the choice of whether to achieve the objective using pump-and-treat, a biobarrier, or natural attenuation has a direct impact on costs (including any external costs associated with the method, such as release of off-gases to the atmosphere [see Chapter 8]), but benefits remain essentially constant.

Even with this limited number of remedial objective options, however, the analysis and the choice of remedial objective become quite complex when mobile groundwater plumes are involved. Figure 10.1 provides a simple visual schema for considering the overall consequences of various remedial objective options under such conditions. A fixed point-source actively introduces contaminants into groundwater at a mass-rate dm/dt. A contaminant such as MtBE can be considered to behave as a conservative solute, moving at the linear advective groundwater velocity (pore velocity) v, as in Equation 10.1:

$$v = \frac{Ki}{n_e} \tag{10.1}$$

where K is the hydraulic conductivity of the medium, i is the groundwater gradient, and n_e is the effective porosity of the medium. Other contaminants, such as benzene and TCE, will degrade biologically over time and are also subject to adsorption onto matrix material.

As time passes, the plume migrates in groundwater, dispersing laterally and transversely due to the effects of chemical diffusion and mechanical mixing. At time t_1, for instance, the plume has migrated only a short distance and is relatively highly concentrated. Only a relatively small volume of aquifer has been impacted, and the first receptor set down-gradient of the source (receptor set 1, or R_1) has not been

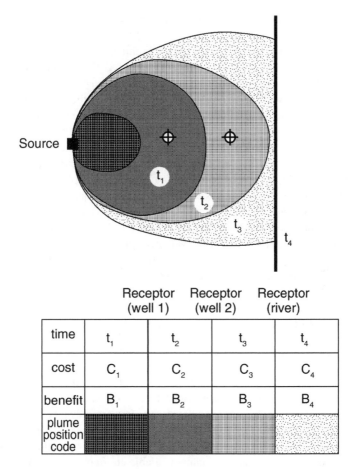

FIGURE 10.1 Schematic of mobile plume affecting receptors over time.

impacted. However, as plume migration continues, receptors will become impacted: first R_1 (wells in this example), then R_2, and finally R_3 (a river). The result is that the number of receptors impacted increases with time. Depending on the behavior of the contaminant solute (degree of attenuation by adsorption, biological and chemical degradation, and dispersion), impacts could also vary with time at a given receptor. Thus, in such a situation, we may need to describe risk as a function of time and space, as in Equation 10.2:

$$Risk = fn(x,t) \tag{10.2}$$

This concept, and its overall potential relationship with costs and benefits, is illustrated in Figure 10.1. As the plume migrates and disperses with time, receptors R_i are impacted at times t_i. At each (x,t) coordinate, a remedial cost C_i and benefit B_i (equal to damage avoided if the remediation took place) would exist. So for the generic case, we see that the costs and benefits of remediation vary not only with time but with the location in space at which we decide to implement our remedial objective.

Within this context, the various options for remedial objective can be seen. The remedial objective could be to prevent impact to the river (R_3). A secondary objective could be to prevent contamination from surpassing a given concentration in receptor wells (R_2). Receptor wells R_1 may already be impacted by the time we discover the problem and the plume has been identified as existing at stage t_2. Note that remedial objectives can be interdependent: achieving one objective might be a prerequisite to achieving another or might substantially help in achieving it. Achieving one objective might automatically mean another is achieved at the same time, and so on. This is a result of the spatial and temporal dynamics of plume movement.

This naturally leads to situations involving remedial objectives that change with time. An initial evaluation could indicate that net benefits are maximized if a certain noncritical receptor could be sacrificed (R_2) and that the situation with respect to a more distant receptor (R_3) could be reevaluated, given that migration times would be expected to be long (tens of years) and attenuation active. At the reevaluation point, remedial technology may have changed, the value of the receptor may have increased, and the degree of attenuation and migration may have turned out to be different from that which was originally predicted. A reanalysis of the costs and benefits could indicate that a change of objectives is warranted.

10.1.3 REMEDIAL APPROACH LEVEL

As discussed in Chapter 3, there are now literally dozens of remediation technologies available to achieve any specific technical outcome. For example, dissolved volatile organic compounds can be removed from pumped groundwater by air-stripping (packed tower, shallow tray, or other configurations), advanced oxidation methods (UV-ozone, hydrogen peroxide, TiO_2-UV, and other systems), granular activated carbon (GAC), biological reactors (many available configurations), and many other methods. Poly-aromatic hydrocarbons (PAHs) can be removed from contaminated soil by excavation and treatment by soil washing, enhanced biological treatment, or chemical means. This is the most detailed level of assessment of remedial costs. Often several different technologies, each designed to achieve a specific technical outcome, will be required to create a system that can accomplish the remedial objective. Because so many technologies exist, which achieve such different technical outcomes at such widely varying costs, explicit cost–benefit analysis could literally require comparison of dozens of cost (technology) options, many of which are designed to achieve very different technical outcomes. The link between the remedial objective and the many pieces of technology available on the market today, is the *remedial approach* (sometimes called *remedial strategy*).

The remedial approach does not focus on technology but on ways of breaking the pollutant risk linkage that causes (or will cause) damage. The list of possible remedial approaches is relatively short: either remove the *source*; eliminate the *pathway*; or protect, move, or manage the *receptor*. Consideration of the remedial approach can be very useful in streamlining the cost–benefit analysis process, because a limited number of approaches must be considered. In this way, the remedial approach provides a link between remedial objectives and the hundreds of technologies available. Also, the degree to which the linkage is broken, the timing of the action, and the spatial location at which the action is taken are all variables that must be considered when choosing the approach. A constraints analysis can be undertaken to help assess which approaches are realistically achievable.

Preliminary, high-level costs can be assigned to each approach that can feasibly achieve the desired objective and compared to the benefits of achieving the objective. This provides a relatively quick strategic analysis of the costs and benefits of remediation, before proceeding to detailed technology evaluation and cost analysis.

10.1.4 REMEDIAL TECHNOLOGY SELECTION LEVEL

The remedial technology selection level involves choosing the most cost-effective way of putting a remedial approach into play to achieve a remedial objective. This requires detailed comparison of capital and operation and maintenance (O&M) costs for technologies over a given project life span. External costs should also be incorporated into the cost analysis. Application of various constraints provides a life-cycle cost analysis. By using the intermediary remedial approach step, an iterative analysis of the project is possible. Remedial technology selection (essentially a least-cost exercise) is used to determine the lowest cost option to achieve a given remedial approach.

10.2 DEVELOPING AN ASSESSMENT FRAMEWORK

10.2.1 REQUIREMENTS

The framework presented in Part III is intended to allow the transparent and consistent assessment of costs and benefits in the remediation of polluted groundwater on a site-specific basis. This chapter presents a detailed framework for decision making. Eventually, this or similar frameworks may be more fully developed into working models.

The following are considered essential to a workable framework:

- Procedure for developing the remedial objectives for groundwater cleanup (the level to which groundwater should be remediated)
- Ability to cope with multiple objectives for a single groundwater contamination problem that generates several very different risks (different SPR linkages)
- Facility for considering explicitly the mobility of groundwater contamination and the fact that risks of exposure to receptors may increase over time
- Capacity to include the wider economic benefits of groundwater in the analysis, even on a semiquantitative level, perhaps through the use of threshold value concepts ("We know that the value of the groundwater is at least this much, based on readily available benefit assessment data, even though we have not included the more difficult-to-measure benefits.")

- Some level of probabilistic analysis of the results of the economic analysis, explicitly recognizing the significant level of uncertainty that exists with many groundwater contamination problems
- Constraints to remediation that are unique to groundwater, including the limitations of present technology in dealing with complex and deep-seated contamination and physical restrictions on access to the aquifer because of surface obstructions and property rights issues (off-site)
- Ability to consider long planning horizons, particularly in situations in which irreversible or widespread contamination exists
- Flexibility to incorporate changing remedial objectives within a long planning horizon

The framework presented in the rest of Part III allows readers to use economic arguments when comparing a number of remedial objective options. This critical decision will, in many ways, dictate which remedial approaches and technologies will be applicable and which will not. As such, the selection of remedial objective will have a profound impact on the cost of remediation. The more stringent and aggressive the objective, the higher the cost is likely to be. Thus, objectives set without at least some reference to the costs and benefits of remediation may be unrealistically aggressive and may not reflect a wider economic optimum. Conversely, if the wider benefits of actions to remediate and protect groundwater are not considered at an early stage and only private benefits are included, the remedial objective may not be ambitious enough, and society may unknowingly incur substantial losses in well-being. Figure 10.2 represents some of these considerations graphically. The cost–benefit framework may thus be seen as a tool for negotiation and consensus-building among the various stakeholders.

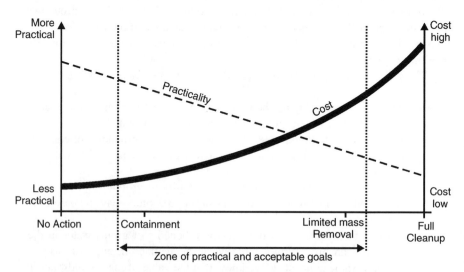

FIGURE 10.2 The relationship between remedial objectives (goals), costs, and practicality of implementation.

10.2.2 FRAMEWORK OVERVIEW

The framework presented herein is designed to allow the user to enter the process at any stage (see Figure 10.3). If a remedial objective has already been determined (by policy, legislation, or consensus), the user moves directly into selecting the least-cost approach and technology. In this case, cost–benefit analysis in the strictest sense is not required, because we are simply finding the least-cost alternative that reaches the set objective. If both objective and approach have been predetermined, the framework assists with selection of the most cost-effective technology or technologies, given an analysis of applicable constraints to remediation.

Remedial objective options are assessed using CBA or partial CBA in concert with multicriteria analysis (incorporating a constraints and policy analysis). Information provided by the risk assessment is a key input in determining both costs and benefits, and thus the analysis can be called *risk–CBA*. Benefits of each objective are assessed, at least as threshold levels (partially monetized). Care is taken to guide the user toward including the full range of likely benefits, including external benefits. Remedial approaches with which the objective can feasibly be met are used to develop indicative costs for optimally reaching the objective. Thus, various possible objectives can be compared using CBA or partial CBA.

Once the remedial objective is set, using the optimal approach, a more detailed analysis of the least-cost technologies with which to implement the approach is undertaken. Because of the importance of time in groundwater remediation, a constraints analysis is applied to life-cycle cost–time functions developed for each technology option under consideration. This is essentially a least-cost analysis exercise.

At this point in the process, an iterative checking of assumptions used in the high-level analysis may be needed. If the detailed technology assessment reveals that a lower-cost implementation of the selected approach is possible, the benefit–cost ratio of the selected objective may also change. If the change is significant

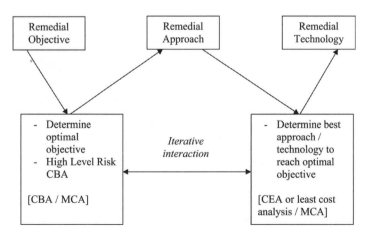

FIGURE 10.3 Economic analysis tools used within the framework.

enough, it could mean that another objective is more economically advantageous. In this, the remedial approach can be seen as the vital link between remedial objective (strategy) and remedial technology (tactics). In practice, it is considered likely that decisions reached using this analysis will be fairly clear-cut on most occasions, especially once users begin to gain experience with the framework.

11 Setting an Optimal Remedial Objective

11.1 INTRODUCTION

This chapter and Chapter 12 lay out a structured framework for using cost–benefit analysis to choose the most beneficial remedial objectives for a given problem and the best ways of reaching those objectives. The framework includes each of the main issues discussed in Part II and in Chapter 10. In practice, only the most high-profile and difficult sites will warrant use of the whole framework. Some steps are more demanding and more difficult to execute than others, and some require information that is not always available. More commonly, certain steps can be skipped, and the analysis can be focused on the critical and readily executed steps. Part IV provides case histories in which the framework is followed to varying degrees. The intent of the framework is to provide a starting point. In practice, users should consider each case individually when deciding how best to use the framework and which steps to use or discard.

A conceptual framework for supporting the selection of an economically suitable remedial objective for groundwater remediation and protection is illustrated in Figure 11.1. This figure is in the form of a flow chart, moving the user through a number of steps toward selection of an economically optimal objective. The process is divided into five main steps, each of which contains a number of substeps.

The intent was to develop a simple method that allows a gradual screening of alternative objectives. Initial screening (Step A1) is based on the results of risk assessment and application of basic policy and constraint criteria. At this stage, a short-list of practical objectives is developed. Alternatively, an objective could be selected immediately at this step, and the user would be directed toward the approach/technology selection step (Step A2).

If several alternative objectives are possible, suitable approaches are identified (Step A2) and the benefits of each are assessed (Step B1). Then the least costly way to achieve those benefits is determined, through comparison of feasible remedial approaches for reaching each objective (Step B2). This leads to conducting a high-level CBA or partial CBA (through inclusion of nonmonetary criteria) to compare approaches for each objective (Step B3) and among objectives (Step C).

11.2 STEP A1: SCREEN OBJECTIVES — DEVELOP SHORT-LIST

This step uses as its primary input the results of the risk assessment (RA) carried out on the groundwater contamination problem. The key source–pathway–receptor (SPR) risk linkages are determined, and the risks associated with each complete

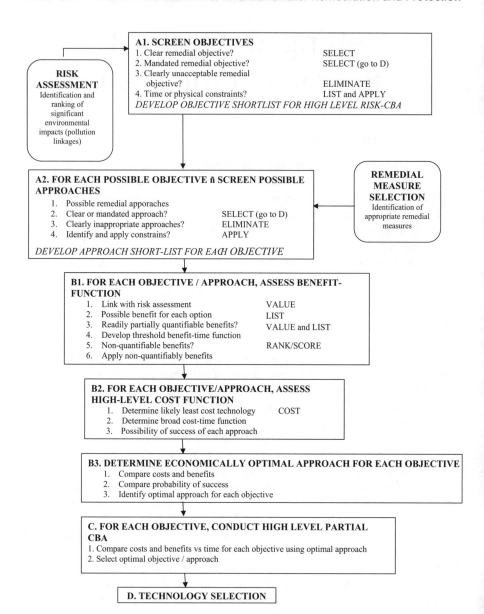

FIGURE 11.1 Determining remedial objective(s) — high-level analysis.

linkage are assessed. Depending on the level to which the RA is completed, the various linkages that exist on each site will be ranked according to the severity of the impact on the receptor, the sensitivity of the receptor, and the probability of the impact occurring.

Whether the environmental RA has been conducted to a qualitative or a quantitative level, some indication of the relative ranking of various SPR linkages can be made. For each completed risk linkage defined in the RA, a remedial objective analysis should be conducted. In many cases, multiple linkages may exist, so there may be several parallel or competing remedial objectives. The ranking of risks completed as part of the RA would be used to provide a preliminary prioritization of objectives. The objectives analysis could potentially be completed for each pollutant linkage, if the linkages were sufficiently different to warrant this. The judgment of the user would be important in this determination.

11.2.1 STEP A1-1: IDENTIFICATION OF A CLEAR REMEDIAL OBJECTIVE OR OBJECTIVES

In many situations, risk assessment will reveal a clearly defined risk that is unacceptable. In these situations, the elimination of that risk may be a clear objective for which no further analysis is needed. Such examples could include spills of particularly toxic substances into groundwater, which are directly affecting public health. Examples of such situations are provided in Table 11.1. In such a situation, direct and immediate action to protect the affected members of the public is required. The remedial objective is selected. A cost–benefit analysis in support of remedial objective selection is not strictly required — the objective is clear and agreed on by all parties.

If an objective is selected, the user would move to Step A2 and develop a short-list of remedial approaches that can reach the remedial objective(s). In developing this framework, there was some debate about whether the remedial objective alone was enough to evaluate benefits, leaving the approach/technology selection as a purely least-cost analysis exercise. However, after we developed several groundwater contamination scenarios, it became clear that the approach (a high-level strategy for the remediation) could have an impact on benefits. Technology selection, however (the way of achieving the approach), did not impact benefits. Thus, the benefit analysis must be conducted for each objective/approach option.

If the objective is not clear and several possible objectives appear reasonable, the user would move to the next step (Table 11.2).

Note that identification of a clear remedial objective may still mean that other objective options need to be examined more closely for their economic viability or that the immediate clear objective would be supplanted in time with a revised objective, depending on circumstances in the future (*temporally changing objectives*).

TABLE 11.1
Criteria for Clear Remedial Objectives: (If Yes, Select)

- Immediate serious impacts to human health

- Immediate serious impacts to safety (explosion or fire hazard)

- Immediate serious impacts to sensitive designated ecosystems or the environment

---

11.2.2 STEP A1-2: IDENTIFICATION OF A MANDATED REMEDIAL OBJECTIVE

In some situations, the results of the RA will indicate to regulators that the public interest or law mandates a clear remedial objective. For instance, in the United States, CERCLA identifies priorities for receptor protection. The U.K. Environment Agency has already developed a policy and tools for the protection of groundwater and prioritization for its remediation.[1] The priorities for protection are, in descending order:

- Groundwater currently being used for potable water supply (Source Protection Zones)
- Unexploited major aquifers
- Unexploited minor aquifers
- Sites located on nonaquifers

This prioritization must be fully considered in light of the seriousness of the pollution threat presented.

As with step A1-1, if this is the case, the mandated objective is selected. Situations in which a mandated objective may exist are listed in Table 11.2. In many ways, this substep is similar and shares common features with A1-1.

If the objective is not clear and several possible objectives appear reasonable, the user would move to the next step (A1-3). If an objective is chosen, the user would move to step A2.

Several considerations of a noneconomic nature will naturally be important to the decision-making process, including public and private policy (e.g., corporate environmental policy) regulations and stakeholder views. Despite the fact that many of these factors may be expressed in economic terms, it is clear that this will not be practical, nor is it likely to be accepted by many of the involved parties. As such, it may be preferable that the remedial objective is determined by first setting a minimum acceptable range of conceptual objectives, based on the rule of law, environ-

---

TABLE 11.2
Criteria for Mandated Remedial Objectives

- Legal requirement to take action to protect sensitive receptor or human health (e.g., CERCLA [U.S.], SDWA [U.S.], Environmental Protection Act 1995 [U.K.])

- Policy requirement to take action to protect sensitive receptor (e.g., SDWA [U.S.], Policy and Practice for the Protection of Groundwater [U.K.])

- Legal requirement to protect critical aquifer or groundwater resource (e.g., SDWA [U.S.], Water Resources Act 1991 [U.K.])

- Direct Presidential or ministerial intervention in setting remedial objectives (e.g., Presidential Directive [U.S.], Government Circular [U.K.])

- Direct requirement to take action under environmental regulations (e.g., CERCLA [U.S.], Groundwater Regulations 1998, Environmental Protection Act 1991 [U.K.])

mental regulations, and the results of risk analysis and stakeholder consultation. At this stage, a benefits analysis could be used to set a minimum and maximum range on the benefits of remediation, considering a range of conceptual remedial approaches. This would provide a range of acceptable objectives, from which a final objective can be selected, based on the input of all stakeholders, results of a preliminary economic analysis, and negotiations between the problem holder and the regulators.

To this is added a high-level constraints/policy analysis, which identifies key constraints that would effectively rule out certain objective options. The key element of this is risk assessment, which provides guidance on which pollutant linkages must be dealt with and in what order of priority. In this, there is an explicit link with environmental risk assessment, where risks are identified and categorized. Policy helps to put the various risks into perspective and provide guidance on prioritization.

11.2.3 STEP A1-3: IDENTIFICATION OF CLEARLY UNACCEPTABLE REMEDIAL OBJECTIVES

The list of possible remedial objective options may be further narrowed, if one has not already been chosen, by removing remedial objective options that are unacceptable for one reason or another. In some situations for example, the objective of reducing impact to a current receptor may not be acceptable, depending on the level of the intended reduction. In these cases, these objectives would be eliminated. A list of criteria for unacceptable remedial objectives is provided in Table 11.3. The same follow-on procedure as described for substeps A1-1 and A1-2 would then be followed.

11.2.4 STEP A1-4: PRELIMINARY CONSTRAINTS ANALYSIS

Constraints to remediation can take many different forms: time pressures, physical limitation on placement of equipment, technological limitations, land ownership, and property access constraints are just some of the issues that could affect the selection of remedial objective. In addition, some objectives, such as removal of all contamination from the aquifer, may be technologically impractical (TI), even if substantial funds were available. An example of this would include complete removal of deep NAPL contamination in fractured rock. Table 11.4 lists some of the constraints that should be assessed in determining whether objective options are generallyfeasible.

TABLE 11.3
Criteria for Clearly Unacceptable Remedial Objective Options

- Risk reduction insufficient to prevent ongoing unacceptable damage to sensitive receptor
- Risk reduction insufficient to prevent future unacceptable damage to sensitive receptor
- Objectives that would clearly contravene government policy, legislation, or rule of law

---

TABLE 11.4
Constraints Analysis

Time Constraints

Are there specific limitations on the duration of remediation?
Are there specific limitations on when remediation needs to be implemented?
Is there a need for immediate action (independent of other considerations)?

Physical Constraints

Are there access restrictions that may affect remedial objectives?
Are there physical restrictions that may affect remedial objectives?
Are there physical constraints that may limit options in management of receptors?
Will the remedial schemes affect operations of a working site?

Technological Constraints

Are there objectives that are technically infeasible/impractical?
Are there objectives that are clearly cost prohibitive (on the grounds that they are unreasonable, rather than that they would cause hardship to the current problem holder)?

This analysis is focused purely on fundamental factors that could render an objective impractical or that would qualify or constrain an objective. Similar, but more detailed and focused constraints analyses would also be performed at the remedial approach analysis stage and at the technology selection stage.

11.2.5   STEP A1-5: OBJECTIVE SHORT-LIST DEVELOPMENT

After the user has proceeded through the preceding substeps, a short-list of generally acceptable and viable remedial objectives will emerge. At any stage in the previous substeps, a single remedial objective may have been selected. If so, this substep is bypassed. A short-list should not include more than two or three objectives for each SPR linkage (or group of similar linkages sharing common elements) being considered.

The user would then proceed directly to Step A2 (Chapter 12) to develop remedial approaches best suited for reaching each of the short-listed objectives.

REFERENCE

1. Environment Agency, *Policy and Practice for the Protection of Groundwater*, Environment Agency, Bristol, 1998.

12 Reaching the Objective

12.1 INTRODUCTION

After a short-list of remedial objectives has been identified for each source–pathway–receptor (SPR) risk linkage (Chapter 11), suitable remedial approaches for achieving each objective are considered. As discussed in Chapter 10, the remedial approach, which focuses on how the SPR linkage is to be broken, is the key level of consideration. As shown in the examples in Part IV, costs and benefits are compared for the various remedial approaches, alone or in combination. For each approach, a least-cost technology is chosen (assumed). In practice, for less complex sites or when remedial objectives are straightforward, cost–benefit analysis (CBA) can start at this stage, skipping the remedial objective steps (step A1).

12.2 STEP A2: DETERMINE APPROACH SHORT-LIST FOR EACH OBJECTIVE

12.2.1 Step A2-1: List Possible Remedial Approaches

As discussed earlier, remedial approaches focus on how the pollutant linkage is to be broken to achieve the specified objective. Benefits can vary, depending on the remedial approach selected. Thus, the benefits part of the economic analysis requires that an objective be coupled with an economically optimal approach. Determining the most economic approach means that alternative approaches must be considered for each objective and the costs and benefits of each evaluated at a high level. Only then can the optimum objective be determined. At this stage, a list of appropriate approaches for each objective is developed, based on the user's experience, knowledge, and use of available guidance.

12.2.2 Step A2-2: Identify Clear or Mandated Approaches

In any given situation, it may be clear to all stakeholders that one particular remedial approach is the clear choice for achieving the objective, either due to the particular physical or environmental considerations at the site or because of specific legal or regulatory requirements for action. In this case, an explicit comparative analysis would not be considered necessary, and that approach would be selected. Possible criteria for selection of an approach are listed in Table 12.1.

If no clear choice emerges, the user proceeds to step A2-3. If a clear choice is found, the user goes to step A2-4, applies constraints to refine the approach, and then moves directly to step B1.

TABLE 12.1
Criteria for Selection of Approach

- Clear technical justification: No other technically feasible approach exists that can meet the objective, or alternatives are obviously less effective or more expensive.

- Clear environmental justification: No other approach achieves the necessary receptor protection required, or alternatives are clearly inferior.

- Clear regulatory direction or mandate.

- Clear legal imperative.

- Clear economic justification: In some cases, a particular approach is by far the most cost-effective way of reaching the objective, and there are no discernible differences in overall benefits.

TABLE 12.2
Criteria for Rejection of Approach

- Clear technical justification: The approach is clearly technically infeasible.

- Clear environmental justification: The approach does not achieve the necessary risk reduction or produces unacceptable side effects or external costs.

- Clear regulatory direction or mandate: The approach is rejected by regulators.

- Clear legal imperative: The approach would create a situation in which statutes were violated.

- Clear economic justification: In some cases, a particular approach is obviously far too expensive compared to other options for further consideration and does not generate benefits with any discernable difference.

12.2.3   STEP A2-3: IDENTIFY CLEARLY INAPPROPRIATE APPROACHES

As with step A2-2, clear reasons often exist why certain remedial approaches are not acceptable, either to one or to many stakeholders. These can be ruled out from further consideration. Table 12.2 provides a list of criteria that could render an approach unsuitable.

Rejected approaches help to narrow the short-list. The user would then proceed to step A2-4.

12.2.4   STEP A2-4: IDENTIFY AND APPLY CONSTRAINTS

As discussed for remedial objectives (step A1), there will always be constraints that will limit how, where, and when an approach can be implemented. These constraints should be identified based on site-specific conditions. Table 12.3 provides a list of typical constraints that may apply in the case of remedial approaches. The list is not exhaustive, and other factors may also limit the implementation of remedial approaches. Then, the constraints should be applied to each remaining approach (or to the previously selected approach) to refine and focus the approach. Key factors

---

TABLE 12.3
Possible Remedial Constraints

Physical
- Physical site restrictions (gas mains, utilities, way-leaves, buildings)
- Land access restrictions (ownership, access agreements, vital facilities)
- Power restrictions

Regulatory
- Noise restrictions
- Emissions restrictions
- Traffic restrictions
- Odor restrictions

Technical
- Hydrogeological constraints, contaminant type considerations, depth constraints
- Dependability constraints

Economic
- Budget restrictions (private) or objectives
- Cost-timing restrictions (cash flow) private
- Tax considerations

Time
- Need for immediate/emergency response action
- Need for objective to be reached in a certain time
- Need to remove capital equipment from site in a certain time
- Planning horizon restrictions

---

would include where the approach would be implemented (physical location) and when (x,t). In some cases, constraints can be used to eliminate an approach from further consideration.

12.2.5 STEP A2-5: DEVELOP APPROACH SHORT-LIST

The previous steps will result in a remedial approach short-list. Each approach would then be developed into a simple conceptual model or plan, detailing where and when the approach would be applied and how the approach would achieve the remedial objective(s). The user would then proceed to step B1 to begin the economic analysis of the approaches, to determine the most economic way of achieving the objective.

12.2.6 STEP A2-6: DETERMINE RELATIONSHIPS BETWEEN MULTIPLE OBJECTIVES

In many situations, there will be several remedial objectives, all of which will be desired outcomes. Often a main, or preliminary, objective can be readily refined, usually based on the results of the risk assessment. Various approachoptions for achieving each objective will have different impacts on each objective. The relationships can be described as falling into the categories shown in Table 12.4.

When several different objective options are being considered, Table 12.5 provides a framework for recording the relationships between various objectives/approaches. The designations described earlier are used in this simple matrix table.

A_{12} is approach option 2 for reaching objective 1, and 0, +, C and A are the designations described in Table 12.4. In this case, for example, approach A_{11} has positive impacts on achieving approaches A_{22} and A_{23}, both of which are contingent on approach A_{12} being successfully implemented.

Take, for example, a situation in which LNAPL exists in groundwater, producing a dissolved-phase plume that is impacting a human receptor and that will impact a river in the near future. The primary objectives will clearly be the protection of the human receptor and elimination of the risk presently being experienced. However, protection of the river may also be desired. Thus, the river protection objective is secondary (based on legal requirements and comparative economic values), both in terms of relative importance and in terms of timing. In this example, the objectives are closely interrelated, especially when considering approach options for achieving each. Removal of the LNAPL plume as an approach will help to achieve both objectives but may not achieve either without some additional containment or an optimistic natural attenuation program. Containment of the plume at a point just up-gradient of the human receptor would also prevent additional contaminant from reaching the river, but the full effect of this measure would not be felt for some time. Containment of the plume at the river would not achieve the primary objective.

12.3 STEP B1: ASSESS HIGH-LEVEL BENEFITS FUNCTION FOR EACH OBJECTIVE/APPROACH

If an objective has already been selected in step A1, the user may wish to assess the economic benefits of the action, despite the fact that, strictly, the justification for conducting CBA has been eliminated by pre-determining the objective.

TABLE 12.4
Possible Relationships between Multiple Objectives

1. No effect (0): The approach under consideration has no effect on another objective or another approach.

2. Positive effect (+): The approach will have a positive effect on the implementation of another approach or objective.

3. Contingent (C): Some approaches will be contingent on successful implementation of another approach. For instance, applying natural attenuation to reach one objective may require that the subsurface source (LNAPL on groundwater in the example) be removed first.

4. Achieves the other goal (A): The approach being considered for one objective actually accomplishes another goal at the same time.

5. Negative effect (–): The approach has a net negative effective on another objective or approach.

TABLE 12.5
Objective/Approach Relationship

Objective		Approach Objective 2		
		A_{21}	A_{22}	$A_{23}...$
Objective 1	A_{11}	0	+	+
	A_{12}	0	C	C
	$A_{13}...$	A	+	+

12.3.1 STEP B1-1: LINK WITH RISK ASSESSMENT

The results of the risk assessment (RA) were used in step A1 to assist in selection and short-listing of remedial objective options. At this stage, the RA is used to provide specific information on identified receptors, the degree of current and anticipated impact, and the relative importance placed on those receptors. As discussed in Chapter 10, remedial objectives are very much *receptor focused*. Receptor information from the RA is catalogued as shown in Table 12.6.

12.3.2 STEP B1-2: LIST POSSIBLE BENEFITS FOR EACH OBJECTIVE/APPROACH

Using the information from the RA from step B1-1, list possible benefits that may accrue from successful realization of the objective, using the particular remediation approach. This stage of the analysis introduces the distinction between a financial analysis of costs and benefits and an economic analysis. These terms are defined in the following list (and discussed in detail in Chapter 5):

- Financial analysis — Analysis of the costs and benefits accruing to the problem holder, from a purely private or internal point of view. When estimating these costs and benefits, market prices are used, including the subsidies or taxes that are included in the market price. Financial analysis does not deal with environmental or other social impacts of an action unless these have a direct implication for the costs and benefits of the problem holder. Table 12.7 presents a selection of benefit categories that can be used in a financial analysis.
- Economic analysis — An analysis of the full costs and benefits of groundwater remediation. This includes costs and benefits to the problem holder and those to the rest of the society. The latter are also known as *external* costs and benefits. This different definition of costs and benefits requires them to be measured differently than in financial analysis. The prices for marketed goods and services that are affected should no longer be market prices but real or shadow prices. Shadow prices are estimated by subtracting (or adding) tax (subsidy) elements from (to) market prices. Subsidies and taxes are referred to as *transfer payments*, and as such their payment

TABLE 12.6
Receptor Information Required from Risk Assessment

- Type of receptor (human, ecosystem, aquifer, natural resource, water body, end-user)
- Receptor description (detailed description of receptor to assist in economic valuation)
- Extent of damage that has already occurred and period of impact
- Consequences of existing damage (lost productivity, health deterioration, reduced land value, loss of water resources, damage to ecosystems)
- Extent of future predicted damage and period of impact
- Anticipated consequences of future predicted damages

TABLE 12.7
Suggested Private Benefits Categories — Financial Analysis

- Increased property value
- Elimination of corporate financial environmental liability
- Elimination of potential for litigation or prosecution (civil and criminal)
- Positive public relations value
- Avoidance of negative public relations
- Protection of a resource used as a key input to an economic process (e.g., water for irrigation or manufacture)
- Avoidance of exposure of on-site personnel to pollutants

does not cause a net change to the costs and benefits faced by the society as a whole. Similarly, litigation expenses or value of bad publicity are also transfer payments. The problem holder's cost for litigation becomes the benefit of the law firm, so the two cancel each other out when an economic analysis is undertaken. Some costs and benefits to the rest of society may not have market prices associated with them. Examples include uncompensated environmental and health effects. However, at least some of these nonmarket effects can still be estimated by using economic valuation techniques. Table 12.8 presents a selection of external benefit categories.

The financial step is useful because it allows problem holders to begin by fully understanding their own position. Then, once the economic analysis is completed, the effects of external costs and benefits can be clearly compared and judged. This is expected to help all stakeholders with decision making. The final model procedure could have a detailed checklist, ensuring that most, if not all, possible benefits were considered. However, the user is presumed to have some degree of knowledge of the basics of economic evaluation and remediation, in much the same way as users

---

TABLE 12.8
Suggested External Benefits Categories — Economic Analysis

- Increased property values that measure the benefit to local people and that may include the health benefits to local people
- Health benefits to visitors to the area
- Recreational benefits to local residents insofar as these are not captures in the change in property values
- Recreational benefits to visitors to the area
- Avoidance of ecological damage not otherwise captured in recreational or property value increase
- Gains in nonuse value
- Gains in option value

---

of currently available risk assessment guidance are expected to have some knowledge of risk assessment, for instance.

For each of the benefits identified, the user would be asked to provide a preliminary assessment of how likely that benefit was to be realized, assuming that the remedial objective was completely met. This could take the form of a simple scale: very likely (VL), likely (L), and not likely (NL). This scale takes into account, on a basic level, the uncertainty associated with this analysis and provides a level of differentiation between benefits that are sure to arise and those that may arise. The degree to which the risk assessment has analyzed predicted impacts will have a bearing on the number of VL benefits. A detailed quantitative risk assessment, for instance, would be expected to produce a more focused list of possible benefits, with a higher proportion of VL responses.

12.3.3 STEP B1-3: IDENTIFY AND VALUE READILY QUANTIFIABLE BENEFITS

For each of the possible benefits identified in step B1-2, the user would identify those benefits that can be readily quantified in monetary terms. In practice, this is likely to include several of the key private benefits (such as land value and economic input values). External benefits are less likely to be readily monetized. The degree to which monetization of benefits occurs would depend on the requirements of the user. For more complex, high-profile, and serious problems, a greater degree of analysis would be warranted. An emphasis should be put, if possible, on monetization of VL benefits rather than NL benefits. However, wherever benefits can be reliably monetized, they should be. Considerations for a rational, objective, and consistent quantification of benefits are listed in Table 12.9. A sample template for benefits assessment is provided in Table 12.10. (This is essentially benefits transfer — see Chapter 4.)

TABLE 12.9
Considerations for Consistent Monetization of Benefits

- Set the planning horizon for evaluation of the project. If the project has a defined end-point, this might be selected. Benefits of damage avoidance, however, will accrue in the long term. Often, 20-year or 50-year horizons are selected, partially from convention and partially due to their reference to one- or two- generation time spans. Another option is to set the planning horizon to reflect the persistence of the pollution.

- Set the discount rate for analysis of flows of money over time, taking into account of the available guidance for public investments (e.g., in the U.K. the social discount rate to be used in public sector investments or investments with external costs and benefits is 3.5%).

- Use actual market data where possible.

- Use benefits transfer to estimate economic value from the existing literature.

- Use revealed or stated economic valuation techniques to undertake original valuation studies for the site of concern.

- Include a confidence interval for estimates, where possible, to reflect risk and uncertainty that surround the various assumptions.

TABLE 12.10
Sample Benefits Quantification Template

Benefit Name	Likelihood	Unit of Benefit (physical)	Unit Value of Benefit (monetized)	PV of Total Benefit	Uncertainty
Name	VL, L, NL	e.g., hectare of redeveloped land	e.g., $/Ha	$	+/- $

12.3.4 STEP B1-4: DEVELOP THRESHOLD BENEFIT–TIME FUNCTION

In practice, in most cases, benefits of remediation will not be completely monetized for reasons discussed in Chapter 4 and Chapter 5. Often, only a small number of the benefits of groundwater remediation can be readily quantified. However, financial costs of remedial action are usually quantified readily and with relative accuracy. Thus, a comparison of costs and that portion of benefits that can be monetized (i.e., a partial CBA) is an effective starting point for economic analysis of options. These partially quantified benefits, or *threshold benefits*, mean that we know that reaching the remedial objective will be "worth at least this much" and probably more. Thus, at this step, the user is asked to develop a threshold benefit function, in which the quantifiable benefits of remediation vary with time. This time-variant behavior of benefits, which can be quite complex, is discussed in Part II and Chapter 10. Again, depending on the complexity and seriousness of the problem, less detailed functions could be developed. From the benefits function, the present value of the benefits would be calculated, based on the set planning horizon and discount rate.

TABLE 12.11
Benefits Apportionment Matrix — Example

	Benefit 1	Benefit 2	Benefit 3	Benefit 4
Approach	Property value increase	Increase in the value of the neighboring properties (blight reduction)	Improvement of river water quality	Improvement of aquifer water quality
S1: Excavate source in sediments	1	1	0.25	0
S2: Remove NAPL in bedrock	0	0	0	0.1
P3: Contain dissolved phase in bedrock — protect well	0	0	0	1
P4: Contain dissolved phase, protect river	0	0	1	0
S1+P3	1	1	0.25	1
S1+S2	1	1	0.25	0.75

Next, the user completes a benefits apportionment matrix (Table 12.11) for each approach being considered. This matrix allows a clear presentation of which of a range of possible remedial benefits is actually realized by the application of a particular approach, and to what degree. In Table 12.11, full realization of a benefit (i.e., avoided damage on a receptor) is indicated by a 1; no realization of the benefit is indicated by a 0. Partial realizations are possible. In determining monetized benefits for each approach, the monetized total value of each benefit is multiplied by its corresponding factor in the matrix. Depending on the remedial approach being considered, different total benefits will accrue.

In the example in Table 12.11, four readily monetizable remedial benefits have been identified at a site where NAPLs are present in sediment and underlying bedrock. NAPLs are acting as a source of dissolved-phase contamination that is discharging to a nearby river via the shallow sediments and to the deeper aquifer. Depending on the remedial approach chosen, benefits can accrue from making the property fit for use and sale, from elimination of blight to the neighborhood through site clean-up, from protection of the river, or from protection of a nearby water supply well completed in the deep aquifer. Excavation and treatment of contaminated sediments at a site (approach S1 in Table 12.11) will result in capturing the private benefit of uplift on the value of the property at sale and the full benefit of removal of blight from neighboring properties. Although these approaches will also make a positive contribution toward restoring the aquifer, they cannot capture these benefits alone. A complementary remedial approach, such as hydraulic containment of dissolved-phase contaminants in the bedrock aquifer, must also be used (approach S1+P3) to capture aquifer benefits. An approach designed to cut the pathway for dissolved phase contamination to the river, however, will yield only river-protection benefits. The site itself would remain in its current state, and the deeper aquifer would not be affected.

So, if the present values of each benefit were established as $2 million for the property value (Benefit 1), $0.5 million for neighboring property value improvement from blight removal (Benefit 2), $0.2 million for river protection (Benefit 3), and $0.75 million for aquifer protection (Benefit 4), using the matrix in Table 12.11, the total benefit for approach S1 would be:

$$1(\$ 2 \text{ million}) + 1(\$0.5 \text{ million}) + 0.25(\$ 0.2 \text{ million}) + 0(\$0.75 \text{million}) = \$ 2.55 \text{ million}$$

In contrast, approach P3, which only protects the well in the bedrock aquifer, would have a total benefit of:

$$0(\$ 2 \text{ million}) + 0(\$0.5 \text{ million}) + 0(\$ 0.2 \text{ million}) + 1(\$0.75 \text{million}) = \$ 0.75 \text{ million}$$

To estimate the net benefit of each approach, the cost of the approach (private and external in case of economic analysis) is deducted from benefits, calculated as above.

12.3.5 STEP B1-5: IDENTIFY AND ASSESS NONQUANTIFIABLE BENEFITS

Step B1-3 identifies benefits that cannot be readily monetized (by elimination). However, as discussed in detail in Part I, some of these benefits are considered by many researchers and workers in the field to be important and, if properly quantified, of significant value. Thus, such benefits must be included in the overall analysis. Table 12.12 provides a sample template which could be used to list and evaluate nonquantifiable benefits of remediation. Benefits are subjectively assessed based on whether they are expected to increase overall benefits, in general or by a negligible (0), marginal (+), substantial (++), or considerable (+++) amount. This can be accomplished, in part, by gauging the expected contribution of the nonquantifiable benefits against the magnitude of overall quantifiable benefits. Clearly, this very subjective step will require care and judgment.

12.3.6 STEP B1-6: APPLY NONQUANTIFIED BENEFITS

In simple terms, the effect of the nonquantifiable benefits on the threshold benefit–time function can be described as increasing the benefits: negligibly, marginally, substantially, or considerably. This distinction would be based on the results of the assessment carried out in step B1-4, and the magnitude of the threshold benefits

TABLE 12.12
Sample Nonquantifiable Benefits Template

Benefit	Likelihood	Relative Benefit Impact	Uncertainty	Comments
Name	VL, L, NL	0, +, ++, +++	High, medium, low	

estimated in step B1-5. Obviously, such a comparison is quite subjective and should be guided by available models and case histories for similar situations in the literature and the experience of the user. If nonmonetized benefits are expected to be considerable, the user may wish to consult with an expert in environmental economics to assist in the determination (some decision-making rules are given in Chapter 5).

Nonquantified benefits would be used to qualify the benefit function. Benefits for a given remedial objective/approach could be qualified as being considerably greater than a certain sum, which would mean that the actual benefits were judged to be some multiple higher than were actually quantified. If the actual benefits could be termed marginally greater than the quantified benefits, we might expect an uplift of a few percent on the quantified value, and so on. In this way, comparisons are made based on a fixed starting point (the threshold benefits) but can be adjusted upward to account for other benefits. Negotiation and incorporation of the views of the various stakeholders (all of whom would tend to rank the likelihood and importance of various benefits differently) could yield an agreed working value or range of values for total benefits.

12.4 STEP B2: ASSESS HIGH-LEVEL COST FUNCTION FOR EACH APPROACH

12.4.1 STEP B2-1: SELECT LIKELY LEAST-COST TECHNOLOGY TO ACHIEVE APPROACH

As described in Chapter 3, there is often a wide choice of available technologies and techniques that can be employed in a remedial approach design. For example, the remedial approach could be containment of a dissolved-phase plume at a given point or line, with full containment to be achieved within a specified time frame, continuing for 20 years (the objective could be to protect a receptor down-gradient of the containment point). Implementing the approach could involve selection of hydraulic containment (pump-and-treat), physical barriers with some level of pumping, semipassive reactive barriers that treat the contaminant as it passes (using air-sparging, reactive walls, biobarriers, or funnel/trench-and-gate systems), reinjection of treated water, and so on. Each technique listed will involve many subcomponent options as well. If pumping of groundwater is required as part of the approach, then decisions are required on the number and type of wells, the number and type of pump, pumping rates and pumping schedules, and so on. Each of these decisions will be based in part on cost, and each choice will have cost implications.

As described previously, benefits of remediation are expected to change very little between different technologies that are used to achieve a given objective/approach. Thus, a high-level economic comparison of remedial objective/approach options requires that, as far as possible, the implementation of the approach is assumed to be on a least-cost basis.

This step, then, provides the conceptual link between the objective/approach level of analysis and the approach/technology level. The user must develop preliminary costs for implementing each approach analyzed. Many procedures and guidance documents are available that describe fundamental criteria for screening and selecting remedial methods, including costs. Thus, if it is assumed that the user has consulted one or more of these, some form of remedial options cost analysis may have been completed.

On this basis, the user is asked to provide a preliminary estimate of least-cost implementation for the approach under consideration. At this stage, the analysis of costs does not have to be detailed, especially considering that benefits (to which these costs will be compared later) are unlikely to be accurate or complete. Basic first-cut cost estimates will suffice. Later in the process, if required, a second iteration can be performed. The basic elements of developing a preliminary remedial cost estimate are shown in Table 12.13. This template can be used to compare quickly the costs of various technology options for the given approach.

The discount rate selected for the cost analysis will have a significant effect on the present value of costs. Often for a complete economic analysis, a social discount rate that takes note of society's preferences for future costs and benefits should be used. In private financial project analysis, on the other hand, much higher discount rates are usually used, reflecting desired corporate internal rates of return and investment payout targets. Note that the discount rate used for calculating the present value of benefits must also be used for costs for the comparison of the two figures to be consistent and valid.

12.4.2 Step B2-2: Determine Broad Cost–Time Function

The results of substep B2-1 provide a broad, high-level cost function that represents what is considered to be the least-cost way of implementing the approach. Development of detailed technology life-cycle cost functions for detailed analysis of remedial approach/technology options is discussed in step D.

12.4.3 Step B2-3: Qualify Probability of Success of Each Approach

Using the selected least-cost technology, a probability of success is assigned to each approach. The USEPA has devised a qualifying system for remedial technologies based on the designations accepted, emerging and experimental. These categories represent broad indications of the degree to which the methods are understood, validated, and field-tested. Definitions of the designation are provided in Table 12.14.

Note that some accepted techniques may be applied in novel or experimental ways, in some conditions. Similarly, there may be high confidence of success for some

TABLE 12.13
Cost Analysis Template for Remedial Approach Analysis

Cost Category	Annual Operation Cost Estimate	Capital Cost Estimate	Total PV Cost over Planning Horizon
One-Time Costs			
Design and engineering			
Mobilization			
Preparatory/enabling Works			
Capital costs: initial			
Capital costs: future modification			
Ongoing Costs			
Operation			
Maintenance			
Treatment			
Monitoring			
Validation			
Mitigation costs for external impacts			
Posttreatment costs			
Decommissioning			
External costs			
Total cost			

emerging or experimental techniques, when applied in specific conditions, environments, or jurisdictions. Thus, these distinctions should be considered in the context of the proposed use of the technology. In addition, it is clear that technological development is rapid in the field of groundwater remediation. Techniques considered emerging at one point in time may soon become generally accepted in the industry. SVE and, to some extent, air-sparging are examples of technologies that have become seen as widely accepted for hydrocarbon clean-ups in the last few years.

12.5 STEP B3: DETERMINE ECONOMICALLY OPTIMAL APPROACH FOR EACH OBJECTIVE

At this point, the user has developed preliminary high-level cost and benefit functions, uncertainty ranges on costs, qualifiers on benefits, and probability of success

TABLE 12.14
Remedial Technology Category Designations

- *Accepted*: *High probability/confidence of success*. Method has been extensively applied under a variety of conditions, is well documented, and has wide acceptance in the industry. The method is being applied to achieve an outcome it is well suited to achieving (e.g., pump-and-treat applied for hydraulic plume control is appropriate, whereas pump-and-treat for aquifer restoration is usually not appropriate). Examples of accepted techniques: pump-and-treat, slurry walls, air-stripping for VOC removal from pumped water.

- *Emerging*: *Moderate probability of success*. Method has been tested at field-scale but is still not fully understood or developed and still requires more detailed field validation. Examples: *in situ* air-sparging, funnel-and-gate, natural attenuation, *in situ* bioremediation.

- *Experimental*: *Low probability of success*. Method is undergoing bench-scale or pilot-scale testing, is not well understood, and is currently the subject of significant study. Examples: phytoremediation, *in situ* surfactant washing.

factors for each approach being considered for the objective. These are compared in a simple partial CBA, qualified by the other factors. The template in Table 12.15 can assist the user in comparing the options and identifying the economically optimal approach. Ranking would be performed by looking at the factors in the following order (also see Chapter 5 for comparison of monetary and nonmonetary benefits and costs):

- PV threshold net benefit (benfits–costs) — If net benefits, using only threshold benefits, are positive, it is a likely indication that the objective/approach is promising. Benefits of various approaches being considered can be compared directly. The larger the net benefit is, the higher ranking the approach.
- Nonquantifiable benefits — If these benefits are considered by the user to be considerable (+++), in rough relation to the threshold benefits yielded by the objective/approach, and may be considered in the same order of magnitude as the threshold benefits, these may even be able to offset a negative net threshold benefit. Negligible or marginal nonquantifiable benefits will have little impact in overcoming a negative net threshold benefit.
- Probability of success — For approaches with similar net benefits (all told), the probability of success would define the relative ranking. However, judgment would be required, and the individual conditions of the site would have to be taken into consideration, when selecting between very similar options.

In the final analysis, it should be remembered that this particular framework is intended first and foremost to provide an *economic analysis* of options. Technical analysis and selection are dealt with elsewhere. On this basis, the costs and benefits should guide the selection within this framework.

TABLE 12.15
Approach Selection Template

Approach	PV Threshold Benefits (X)	PV Costs (Y)	PV Threshold Net Benefit	Nonquantifiable Benefits	Probability of Success	Ranking
Approach 1	From B1-4	From B2-2	(x–y)	From B1-5	From B2-3	
Approach 2						
Approach 3						

12.6 STEP C: CONDUCT HIGH-LEVEL PARTIAL CBA ON EACH OBJECTIVE OPTION

At this stage, an economically optimal approach has been defined for each objective being considered. In a similar fashion to the analysis presented in step B3, the costs and benefits of each objective/approach are compared, and the objective that results in the highest net benefit is selected. If required, the user can now go on to complete a more detailed cost analysis on the chosen approach to confirm the choice of technology or technologies for implementing the approach. If this is required, the user would proceed to step D.

12.7 STEP D: TECHNOLOGY SELECTION PROCESS

Once the objective and the constraints for the remediation have been determined (from steps A through C) and the decision to remediate has been made, attention turns to the most effective technical means of implementing the approach that achieves the objectives. A wide variety of remedial technologies have been developed for groundwater over the past 15 years. Different methods are suitable for different situations and involve different levels of cost. This step in the framework provides simple guidance on a more detailed cost-effectiveness analysis of the technology options.

It is important to note here that remedial costs for groundwater are relatively well understood and documented, especially for the established and emerging technologies. Information on capital and operating and maintenance costs, as well as performance over time and expected lifetimes, are readily available in the literature from organizations such as the USEPA and API and directly from vendors. Also, there is considerable knowledge and experience in the consulting market and within industry regarding costs for typical groundwater remediation solutions. This step can be used in either of two ways:

- The user has completed the high-level objective/approach analysis (steps A through C), which has included a preliminary assumption about least-cost technology implementation of the chosen approach. In this case, the user applies step D to confirm and refine the technology selection and to look at more detailed subcomponent options of the technique, in an attempt to reduce costs further. In this, the user may find that another technology, not originally considered, is actually the least-cost solution. In either of these cases, if the cost used in the high-level analysis is substantially different from the cost identified in the detailed analysis (step D), the user could decide to return to step B and complete another iteration. In some situations, a different objective/approach could be selected. This process is essentially an iterative loop, in which preliminary assumptions are checked and revisited to ensure that the final analysis is accurate. The degree to which iteration proceeds is dependent on the level of detail warranted by the project or problem.

- The user has already identified an objective and one or a limited number of approaches. A life-cycle cost curve can be developed for each approach/technology being considered, constraints applied, and an optimal solution selected.

12.7.1 STEP D1: TECHNOLOGY SHORT-LIST

A list of applicable technologies with which to implement the desired objective/approach would then be developed. In the U.K., the CLR-11 model procedure adequately covers this aspect, including charts that aid in selecting appropriate technologies for the conditions and type of contaminant and a series of detailed technology summary sheets.[1] Further guidance can be found in USEPA and NATO/CCMS publications and a wealth of other USEPA documents (consult www.epa.gov).[2-4] Therefore, this step is not covered further here.

12.7.2 STEP D2: DETAILED CONSTRAINTS ANALYSIS

At this stage, the user is invited to consider revisiting the constraints that may apply to the project but in a much more detailed, focused way, which applies to the individual techniques being considered.

12.7.3 STEP D3: DEVELOP TECHNOLOGY LIFE-CYCLE COST CURVES

Cost and time are perhaps the two most important factors when deciding between remedial solutions that will achieve the same objective. Life-cycle cost functions can be developed for each remedial option being considered. All of the various components of each solution can be costed, and the costs assigned to various stages in the remedial program. Thus, the potentially significant differences between various solutions can be rationalized into a single function, assuming that all will achieve the set objective. Comparing these cost–time functions allows solutions to be compared rationally, and the differences between them assessed.

The advantages of this method are:

- Ensures that alternatives that are designed to achieve the same objective are compared. As discussed in Chapter 3, solutions are often inappropriately compared.
- Uncertainty envelopes on cost and time can be included in the analysis.
- External costs of remediation can and should be included in the analysis.
- Temporal nature of many groundwater remediation solutions can be explicitly accommodated in the analysis, allowing for staged capital investment, intermittent operation, operating and maintenance, and capital replacement schedules to be included.
- Cost and time constraints can be fitted to reveal a range of feasible acceptable solutions.

The cost analysis template provided in Table 12.13 can be adapted for the more detailed analysis and used to assist in developing the life-cycle cost curves by explicitly listing expected costs per year, for each year of the program. In a simple spreadsheet, this could be used to develop cost curves quickly.

A typical life-cycle cost curve is shown in Figure 12.1. The initial capital cost reflects what is required to purchase and commission the remedial system or effect the remedial works (such as subsurface source removal). Capital costs may also be required in subsequent years, depending on the nature and scale of the project. Operation and maintenance (O&M) and monitoring costs may also be required for systems that operate over time. Costs of monitoring remedial progress will almost certainly be required.

Capital-intensive methods (such as aggressive subsurface source removal through excavation or aggressive enhanced steam or surfactant flushing) are typically characterized by cost curves that are initially very steep but reach the remedial objectives relatively quickly and are thus short in duration (steep curve in Figure 12.2). In contrast, passive *in situ* methods (such as a system designed to achieve containment of groundwater contamination and gradual mass removal) may have smaller initial capital costs but significant annual O&M requirements, and may require a considerable period of time to reach a specified remedial objective (flat curve in Figure 12.2). This technique can be used to compare technologies being considered for the selected approach/objective identified in step C.

12.7.4 STEP D4: EVALUATION OF ALTERNATIVES

Life-cycle cost curves can be generated for a number of remedial technologies (or technology combinations) that will achieve the specified objective. By applying the hard constraints of cost (equivalent to the threshold benefit defined in step B1-4) and time (the point at which the objective must be reached) to the alternatives, an envelope of acceptable solutions is described, as shown in Figure 12.3. Remedial approaches/technologies whose end-points fall within the envelope are acceptable. The final choice between two or more acceptable solutions would be based on a comparison of total overall cost and the problem holder's requirements for cost

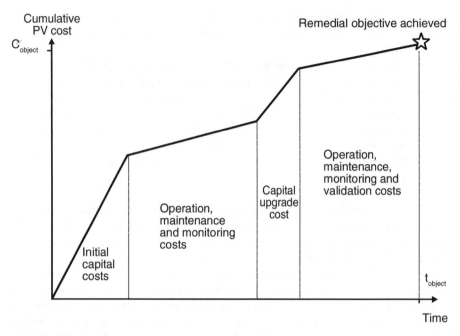

FIGURE 12.1 Typical remedial technology life-cycle cost curve.

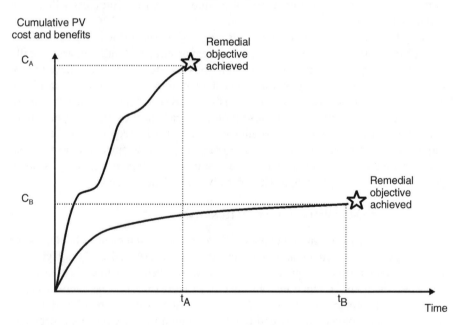

FIGURE 12.2 Comparison of life-cycle cost curves for two technologies.

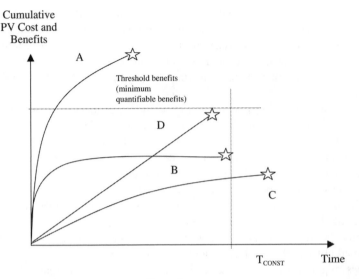

FIGURE 12.3 Applying constraints to life-cycle cost curves.

allocation over time. In Figure 12.3, alternatives A and C fall outside the constraint envelope and are not acceptable (even though they are capable of achieving the objectives of the project). Alternatives B and D are acceptable solutions. Note that despite a higher initial capital cost, alternative B has lower O&M costs and ends up as the least-cost solution. Alternative D would be chosen only if the problem holder wished to spend less initially and were willing to provide for higher subsequent outlays.

In this way, we can see that step D of this framework focuses more on decisions that are made by the problem holder, who is ultimately responsible for bearing the financial burden of the remediation and will most often be in control of the actual implementation. Steps A through C provide for a wider exploration of the economics of the remediation and explicitly incorporate into the analysis the views of the widest possible range of stakeholders.

12.8 DISCUSSION

This framework allows a number of objective options to be evaluated by considering, first, reasonable approaches for meeting the objective through a high-level partial CBA analysis. The optimal approaches for each objective are then compared and an objective identified. The user may then proceed to a detailed analysis of costs and constraints of the technologies used to implement the chosen approach.

Thus, formal CBA (albeit usually only partial CBA) comes into use at the strategic objective-setting and approach-selection levels. Here, the benefits of achieving a certain objective are assessed in a preliminary benefits analysis. To the extent that benefits can be monetized, they are. Where benefits cannot be monetized but clearly exist, a multicriteria analysis (MCA) approach is added to create a benefits threshold approach: "We know that achieving this objective will result in benefits of at least this much (monetized) and that several key additional benefits will be accrued (nonmonetized)." Technology selection is a least-cost analysis exercise: What is the least-cost way of achieving the objective within the identified constraints?

In a similar way, steps A through C provide a framework in which all stakeholders can have input into the economic analysis. This is where wider economic issues (external benefits and costs) are applied and considered and the fundamental direction (objective) of the remediation decided. Step D focuses on the details of implementation and is much more the realm of the problem holder, who is responsible for ensuring the objective is met and ultimately must bear most or all of the financial burden. In step D, the problem holder exercises his available options to produce the least-cost solution that reaches the set objectives.

Thus, the framework allows analysis at the remedial objective/approach level or the remedial technology selection level. This explicitly recognizes that many technologies might be involved in executing a given remedial approach. For example, plume containment to protect a receptor might be achieved by hydraulic containment using pump-and-treat (technology), which in turn could involve different numbers of wells of different designs, equipped with different types of pumping and control systems, using one or more of a number of different water treatment options and a number of available options for final water disposal technology (technology suboptions). A cost-effectiveness analysis for the use of different technologies that fulfills one part of the overall approach can also be very valuable. However, this level of detail is probably of most benefit to the problem holder and less valuable to regulators and other stakeholders, who are more interested in the external costs and benefits. Regulators would probably be satisfied with an analysis to the remedial approach level, allowing the problem holder relatively free rein in selecting the component technologies that make up the approach. This decision would likely be contingent on satisfying regulators that the selected technologies will meet the objectives and that they will be operated in an appropriate manner, subject to relevant authorizations.

REFERENCES

1. Environment Agency and DETR, CLR-11 *Handbook of Model Procedures for the Management of Contaminated Land* (draft), Environment Agency, Bristol, 1999.
2. NATO/CCMS, *Evaluation of Demonstrated and Emerging Technologies for the Treatment and Clean-up of Contaminated Land and Groundwater (Phase II)*, USEPA Ref. EPA S42-R-98-001a and EPA 542-R-98-001c, Washington, DC, 1998.
3. USEPA (U.S. Environmental Protection Agency), *Superfund Innovative Technology Evaluation Program*, EPA/540/R-97/502, Washington, DC, 1996.
4. Environment Agency, *A Methodology to Derive Groundwater Clean-up Standards*, R&D Technical Report P12, by WRc plc., Environment Agency, Bristol, 1996.

Part IV

Case Studies

13 Remediation of a Manufactured Gas Plant Site in the United Kingdom

This analysis follows directly the detailed analysis methodology provided in Part III, to illustrate its full application. In reality, most sites will only require the user to follow parts of the methodology or to simplify certain steps. In the following chapters, more streamlined example analyses are presented.

13.1 OVERVIEW

A hybrid site has been developed based on field data gathered from a number of industrial sites in the East Midlands, U.K. Studying a hybrid site has enabled real data and a variety of existing and potential groundwater remediation issues to be investigated. This hybrid site was originally used in the U.K. Environment Agency report on application of CBA to groundwater contamination problems.[1] Field data relating to site location, history, geology, hydrology, hydraulic properties, and contaminant situation have been included in the hybrid site. A number of source–pathway–receptor (SPR) linkages are considered.

13.1.1 SITE DESCRIPTION

The hybrid site is a former gasworks, located close to a town center in the U.K., and covers an area of approximately 2 hectares (ha). The site is situated in a shallow valley associated with a river that runs adjacent to the western site boundary. Topography slopes gently downwards toward the northeast, following the river's flow direction.

Industrial development of the site commenced in the early 1900s, and processing facilities changed location, design, and size until the 1980s, when industrial activities ceased. The site was subsequently redeveloped as a commercial office complex. Since that time, a number of site investigations have been conducted to assess the contaminant situation and some soil remediation work has been completed.

A mixture of industrial, commercial, and residential properties surrounds the site. To the north, a garage is built just outside the site entrance. A former industrial site to the south and east is being redeveloped as a commercial office facility. To the west, across the river, are houses.

13.1.2 GEOLOGY

A generalized geological sequence has been developed based on actual site investigation data from the component sites and is presented in Table 13.1.

13.1.2.1 Made Ground

Hard surfacing is found over most of the site (> 90%) and is either tarmac or reinforced concrete on a base of crushed stone. In addition, there are numerous buried hard layers distributed throughout the site. Below the hard surfaced layer, the made ground is a highly variable mixture of crushed brick, stone, and red shale, with occasional coke, spent oxide, and wood. Variable amounts of sand and soil are present as infill between the coarse fragments.

13.1.2.2 Superficial Deposits

The superficial deposits underlying the made ground are alluvial clayey sands with occasional thin lenses of quartzite pebbles and cobbles. These sands lie on the soft weathered surface of the underlying Sherwood sandstone group.

13.1.2.3 Sherwood Sandstone Group

The basal part of the Sherwood sandstone group underlies the entire site. The formation consists of a series of interbedded fine to medium sandstones and fine-grained silty sandstones. The bedding is subhorizontal and closely spaced. The majority of fractures are subhorizontal or inclined at an angle of 30 to 45 degrees. Vertical fractures are rare.

13.1.2.4 Middle Permian Marl

The boundary between the Sherwood sandstone group and the lower magnesian limestone is marked by a 1.5 to 2.5 m thick, silty calcareous clay (marl). This marl is stiff but plastic and contains very few fractures or bedding surfaces.

TABLE 13.1
Generalized Site Geology

Material	Thickness (m)
Made ground: Tarmac/reinforced concrete	0.3
Made ground: Variable loose mixtures of medium-to-coarse crushed stone, brick, coke, and shale, with some sand and soil	2.7
Sand: Clayey sand	2
Sherwood sandstone group: Fine- to medium-grained sandstones, with interbedded siltstones; occasional fractures	25
Middle permian marl: Stiff calcareous silty clay; very poorly fractured	2.5
Lower magnesian limestone: Calcareous sandstone and limestone	To depth

---

13.1.2.5 Lower Magnesian Limestone

The lower magnesian limestone consists of calcareous sandstones and limestone, with variable fracturing.

13.1.3 HYDROGEOLOGY

During the investigation, groundwater was found to be present in the made ground and superficial deposits, the Sherwood sandstone, and the lower magnesian limestone. The interpreted groundwater potentiometric surface and groundwater flow directions in the Sherwood sandstone are presented in the appendix to this chapter. The hydrogeological conditions and hydraulic properties for each unit are discussed in the following sections.

13.1.3.1 Made Ground and Superficial Deposits

13.1.3.1.1 Hydraulic Conductivity
Slug tests, performed in monitoring wells completed in the superficial deposits, gave an estimated hydraulic conductivity of 1.2×10^{-7} m/s (0.01 m/day), a value typical for clay-rich sands. Hydraulic conductivity could not be estimated for the made ground using this technique. However, from observations of groundwater flow into excavations, it would appear that the hydraulic conductivity of this unit is highly variable, depending on the nature of the material, and may be as high as 10^{-4} m/s (10 m/day) where the fill material is coarse.

13.1.3.1.2 Groundwater Potentiometric Surface and
         Flow Direction
In general, groundwater was found at between 1 and 2.5 m below ground level (m bgl). In some cases, groundwater was found to be "held" in buried structures above the level of groundwater outside of the structures. Groundwater levels suggest that groundwater within the superficial deposits flows toward the west and the river. Groundwater within the made ground (other than held water) is likely to be in hydraulic continuity with the superficial deposits and is therefore also likely to flow and discharge to the river (generally to the west).

13.1.3.2 Sherwood Sandstone

13.1.3.2.1 Hydraulic Conductivity
Estimates of bulk hydraulic conductivity vary from 5×10^{-6} m/s (0.4 m/day) to 4×10^{-5} m/s (3 m/day). Lower values from silty layers have only a localized effect. Comparison of values derived from pumping tests and from laboratory measurements on matrix samples suggests that the fractures make only a small contribution to bulk transmissivity but may provide important controls on contaminant migration.

13.1.3.2.2 Groundwater Potentiometric Surface and
         Flow Direction
Interpretation of monitoring data suggests that groundwater flow is predominantly south to north down the river valley with a component of flow toward and into the

river. The hydraulic gradient varies from 0.004 to 0.017. In general, groundwater levels measured in the Sherwood sandstone monitoring wells are similar to those in the neighboring shallow wells. This suggests that groundwater in the superficial deposits and made ground (other than held water) is in hydraulic continuity with that in the underlying Sherwood sandstone.

13.1.3.2.3   Groundwater Flux and Advective Velocity

The groundwater flux within the Sherwood sandstone past the site can be estimated using Darcy's law in Equation 13.1:

$$Q = KiWb \qquad (13.1)$$

where Q is groundwater flux through the site (m³/day), K is hydraulic conductivity (m/day), i is hydraulic gradient, W is the width of site normal to flow direction (m), and b is the saturated thickness of the aquifer (m).

The hydraulic conductivity of the Sherwood sandstone is assumed to vary from 5×10^{-6} to 4×10^{-5} m/s (0.4 to 3 m/day). The thickness of the saturated aquifer is 25 m. The hydraulic gradient is 0.004 to 0.017. The width of site normal to flow direction is 70 m. The range of values for groundwater flux through this aquifer beneath the site is therefore 3 m³/day to 103 m³/day. Average linear groundwater velocity is estimated by Equation 13.2:

$$v = \frac{K_i}{n_e} \qquad (13.2)$$

where v is average groundwater velocity and n_e is effective porosity.

Measured total porosity values for the Sherwood sandstone average 0.25, thus an effective porosity of 0.2 is reasonable. This gives average linear velocities between 9×10^{-3} and 0.3 m/day.

13.1.3.3   Lower Magnesian Limestone

Site investigations and subsequent groundwater quality monitoring indicate that the lower magnesian limestone (LML) has not been impacted by site-related contaminants. On the basis of geological maps and borehole logs, the low-permeability middle permian marl is interpreted to be an effective aquitard confining the LML. It is assumed that there are no significant pathways for contaminant migration from the site to the LML. Therefore, no further consideration will be given to this formation.

13.1.3.4   Hydrology

The site receives approximately 700 mm of precipitation per year. Because the site is predominantly hard-covered, vertical recharge below the site will be relatively low. Most recharge to material beneath the site, however, will be via lateral inflows from the surrounding higher ground. The river adjacent to the western boundary of the site currently has a water-quality classification of E (poor). The upstream catchment area is predominantly hard-covered, and flow rate in the river responds rapidly to rainfall events due to surface water runoff. Baseflow conditions indicate that a small component of flow is due to groundwater recharge (likely from both superficial deposits and Sherwood sandstone).

13.2 ASSESSMENT OF CONTAMINATION

13.2.1 GROUNDWATER

An area of heavily contaminated soil has been identified in a central location of the site. This material (approximately 4000 m³) is identified as a source of groundwater contamination and extends to a depth of about 5 m in the made ground and superficial deposits, extending to the interface with the Sherwood sandstone. Contaminants include both organic (predominantly hydrocarbons) and inorganic (ammonium, iron, and sulfate) compounds. Contaminated material includes coal tar compounds with a specific density slightly greater than 1.0. These products have migrated as non-aqueous phase liquid (NAPL) into fractures of the Sherwood sandstone. Site investigation data indicate that NAPL exists in fractures to the base of the Sherwood sandstone. The NAPL acts as a secondary source of groundwater contamination. Across the site as a whole, it is estimated that approximately 8000 m³ of contaminated superficial deposits exist, with varying levels of contamination from coal tar, hydrocarbons, and other inorganic gasworks contaminants.

Groundwater monitoring has identified plumes of dissolved-phase contaminants in both the shallow (made ground/superficial deposits) and deeper (Sherwood sandstone) groundwater. Data indicate that the contaminant plumes have reached the river. Monitoring data suggest that currently there is no significant degradation in river water quality as a result of contaminated groundwater from the site discharging to the river.

13.2.2 SOIL AND VAPOR

As discussed previously, an area of significant soil contamination has been identified on site. Theoretical calculations have been made to assess the concentration of contaminants that may exist in vapor phase below the site and also how the vapor may migrate. Some researchers have indicated that vapor-phase plumes resulting from free-phase contamination can have a significant impact on groundwater quality.[2] This mechanism has been reviewed and is not considered significant in this instance due to the generally low vapor pressures of the organic contaminants involved. This has been confirmed by shallow soil vapors on the site and at the site boundaries in areas of heaviest contamination.

13.3 HYBRID SITE RISK ASSESSMENT

13.3.1 PRELIMINARY QUALITATIVE RISK ASSESSMENT

Semiquantitative risk assessments were conducted for each of the real sites that were used to create the hybrid site. The risk assessment identified 11 source–pathway–receptor (SPR) linkages that exist at the site. These are listed in Table 13.2.

TABLE 13.2
Risk Linkages and Remedial Objective Options

	Receptor	Pathway	Source	Probability of Impact
1A	River	NAPL via superficial deposits	NAPL in superficial deposits	High
1B	River	Aqueous phase via superficial deposits	NAPL in superficial deposits	High
2	River	Aqueous phase via bedrock	NAPL in bedrock	Low
3A	Surface water user	NAPL via superficial deposits	NAPL in superficial deposits	Low
3B	Surface water user	Aqueous phase via superficial deposits	NAPL in superficial deposits	Medium
4	Surface water user	Aqueous phase via bedrock	NAPL in bedrock	Low
5	Aquifer and users	Aqueous phase via bedrock	NAPL in bedrock	High
6A	Residents off site	Vapor from NAPL in superficial deposits	NAPL in superficial deposits	Low
6B	Residents off site	Vapor from NAPL in bedrock	NAPL in bedrock	Low
7A	Workers on and off site	Vapor from NAPL in superficial deposits	NAPL in superficial deposits	Medium
7B	Workers on and off site	Vapor from NAPL in bedrock	NAPL in bedrock	Medium

13.3.1.1 Groundwater Modeling

Groundwater modeling was used to develop a more quantitative analysis of the impact of contaminant transport via groundwater to potential receptors. In later sections of this chapter, quantitative impacts to receptors are used as a basis for monetization of the damage avoided by taking remedial action.

13.3.1.2 Model Aims

The aims of this work were to develop a contaminant transport model of the hybrid site and to test the effects of various scenarios on the risk to potential receptors. Various scenarios were modeled, based on a calibrated version of the groundwater flow model for the Sherwood sandstone in the site vicinity. Sensitivity analyses were conducted for variations in hydraulic conductivity, effective porosity, and thickness of the sandstone aquifer; the effects of hard-covering the site; and various nearby abstraction scenarios.

A detailed description of the groundwater model and modeling results is provided in the appendix to this chapter.

13.3.1.3 Model Results

The base case assumes a stable plume at the present time (time zero), based on field observations and the results of modeling. Benzene is used as an indicator of dissolved-phase contamination in bedrock. Benzene has been modeled with a K_{oc} of 0.026 m³/kg and a biodegradation half-life of 231 days (based on field observations). On this basis, the contaminant plume in bedrock is modeled to have reached equilibrium and is no longer expanding.

The mass flux of benzene to the river reaches a steady-state value of 3.5 g/day after 3,000 days (appendix to this chapter). The impact on the river will depend on river flow: higher flows mean greater dilution of the benzene being discharged to the river. For a river flow rate of 10,000 m³/day (within the typical range for a river 3 m wide), the benzene concentration within the river will be 0.35 µg/l. This assumes nondetectable concentration of benzene as background river water quality. The environmental quality standard (EQS) for benzene in fresh water is 30 µg/l.

A PWS (public water supply) well abstracting 5000 m³/day was modeled, using the calibrated groundwater flow and transport model (appendix to this chapter). Reverse particle tracking was used to define the capture zone of the well. The model was run 14 times, varying the position of the PWS, in order to determine the area of aquifer within which a PWS would capture the plume. This yields an unusable area of 11.71 million m², corresponding to a total potential yield of 4600 m³/day, which corresponds to the amount of recharge that falls on this area. More detailed descriptions of model assumptions and results are provided in the appendix to this chapter.

13.3.1.4 Vapor Modeling — Aqueous Phase Partitioning

Although vapor migration is not directly related to groundwater contamination, economic analysis of groundwater remediation requires that all aspects of site contamination be included in the analysis.

A simple model of expected vapor concentrations generated at the site from dissolved-phase concentrations of the main organic contaminants in groundwater is provided by considering Henry's law (Equation 13.3):

$$H_L = \frac{P_x}{C_x} \qquad (13.3)$$

where H_L is Henry's law constant (atm. moles/m³), P_x is partial pressure of the gas (atmospheres at particular temperature), and C_x is equilibrium concentration of the gas in solution in groundwater (moles/m³).

Aqueous phase concentrations were taken as the maximum modeled concentration at the site boundary. The partial pressure was then converted to a concentration using Raoult's law. The results are shown in Table 13.3.

This is a conservative approach and takes no account of diffusion from the water table through the soil to the ground surface. The results represent the maximum concentrations that can be generated above the contaminated groundwater at the groundwater surface. The results indicate that equilibrium partitioning of volatile organics from the dissolved phase into the vapor phase is not expected to produce any significant risks to identified receptors. A more detailed discussion of this modeling is provided in the appendix to this chapter. The findings of the modeling were confirmed at the site by conducting a field vapor survey. This survey indicated

TABLE 13.3
Predicted Vapor Phase Concentrations

	Maximum Dissolved Phase Concentration at Site Boundary (mg/l)	Partial Pressure above GW at Site Boundary	Concentration in Unsaturated Zone Immediately above Groundwater (ppm)
Benzene	0.0004	1.73E-07	5.65E-07
Ammonia	36	9.87E-06	7.02E-06
Naphthalene	0.15	8.46E-06	4.53E-05

no significant concentrations of organic vapors within 2 m of the surface in any of the most contaminated areas.

13.3.1.5   Vapor Modeling — NAPL Phase Partitioning

A more important risk-generating mechanism at the site is the generation of vapors from NAPL contained in the unsaturated zone superficial deposits. Site reports indicate clearly that pure-phase liquid organic coal tar is present near the surface in the made ground and the superficial deposits. Using Farmer's model, expected subsurface vadose zone organic vapor concentrations can be estimated.

In this case, generation of significant equilibrium concentrations of vapor is expected in on-site areas where substantial NAPL exists in unsaturated made ground and superficial deposits. Unacceptable risks to on-site workers exposed to vapors would exist during excavation activities on site into areas where shallow NAPL accumulations exist. If the surface remains unbroken, vapors are not predicted to pose an unacceptable risk to workers on site.

NAPL is not expected to have migrated off site within the superficial deposits in the vadose zone, so vapors are not expected to pose a significant hazard to residents or workers off site. These results have been confirmed, as discussed earlier, with an on-site soil vapor survey, which did not detect elevated concentrations of organics in the surficial deposits.

13.4   APPLICATION OF CBA TO EXAMPLE SITE

In this section, the hybrid site is used to illustrate the step-by-step application of the framework methodology described in Part III. As required, the analysis is followed for each of the 11 completed SPR linkages identified in the risk assessment (Table 13.2).

13.4.1   Step A1: Development of Remedial Objective Short-List

13.4.1.1   Clear, Mandated, or Unacceptable Objectives

As presented in Part III, the first steps involve identification of clear, mandated, and unacceptable remedial objectives for dealing with each of the SPR linkages (steps

A1-1 through A1-3). As per the protocol described in Part III, the number of possible remedial objectives is limited to the categories in Table 13.4.

In many situations, risk assessment will reveal a clearly defined risk that is unacceptable. In these situations, the elimination of that risk may be a clear objective, for which no further analysis is needed. Such examples could include spills of particularly toxic substances into groundwater that are directly affecting public health. In this example, as shown in Table 13.5, clear remedial objectives are identified for 5 of the 11 SPR linkages. For off-site residents and workers impacted by vapors from groundwater contamination (linkages 6A, 6B, 7A, and 7B), the receptors must be protected. In these situations, no further objective analysis is required. The user moves directly to the selection of the appropriate remedial approach to reach these objectives.

If the objective is not clear, however, and several possible objectives appear reasonable, the user would move to the next step (A1-2). In some situations, the results of the risk assessment will indicate to regulators that the public interest or law mandates a clear remedial objective. This falls within the realm of government policy and law. In Europe and the U.K., regulators have already developed policies and tools for the protection of groundwater and prioritization for its remediation.[3] The priorities for protection are, in descending order:

- Groundwater currently being used for potable water supply (Source Protection Zones)
- Unexploited major aquifers
- Unexploited minor aquifers
- Sites located on nonaquifers

In this example, the clear objectives identified earlier would also be the likely ones mandated by the regulatory bodies.

If the objective is still not clear and several possible objectives appear reasonable, the user moves to the next step. The list of possible objectives may be further narrowed, if an objective has not already been chosen, by removing from consideration objectives that are clearly unacceptable. In Table 13.5, several unacceptable objectives are removed from consideration from each of the 11 SPR linkages. For instance, for linkages 1A and 1B (movement of NAPL to the river), 3A and 3B

TABLE 13.4
Remedial Objective Categories

A	Protect potential receptor
B	Reduce impact to current receptor
C	Eliminate impact to current receptor
D	Remove all contamination (source)
E	Remove some of the contamination (to set level)
F	No action

TABLE 13.5
Remedial Objective Analysis Matrix

	SPR Linkage			Step A1-1	Step A1-2	Step A1-3		Objective
	Receptor	Pathway	Source	Clear Objective	Mandated Objective	Unacceptable Objective	Constraints	Short-List
1A	River	NAPL via surficials	NAPL in surficials	C[2]	C[1]	F	On-site action only	C, D, E
1B	River	Aqueous phase via surficials	NAPL in surficials	–	–	F	On-site action only	B, C, D, E
2	River	Aqueous phase via bedrock	NAPL in bedrock	–	–	F, D[2]	On-site action only Technological limit on NAPL removal in BR	B, C
3A	Surface water user	NAPL via surficials	NAPL in surficials	–	–	F	On-site action only	A, D, E
3B	Surface water user	Aqueous phase via surficials	NAPL in surficials	–	–	F	On-site action only	A,D,E
4	Surface water user	Aqueous phase via bedrock	NAPL in bedrock	–	–	F, D[2]	On-site action only[2]	A, E
5	Groundwater Aquifer + users	Aqueous phase in bedrock	NAPL in bedrock	–	–	D[2], C[3]	On-site action only[2,3]	A, B, E, F
6A	Residents off site	Vapors from NAPL in surficials	NAPL in surficials	A[4]	A[4]	F	Limited off-site action possible	A, D, E

6B	Residents off site	Vapors from NAPL in bedrock	NAPL in bedrock	A[4]	A[4]	F,D[2]	Limited off-site action possible	A, E
7A	Workers off-site and on site	Vapors from NAPL in surficials	NAPL in surficials	A[5]	A[5]	F	Limited off-site action possible	A, D, E
7B	Workers off-site and on site	Vapors from NAPL in bedrock	NAPL in bedrock	A[5]	A[5]	F, D[2]	Limited off-site action possible	A, E

Notes:

[1] Assume EA requires no NAPL discharge to river

[2] Technological Infeasibility (TI) of NAPL removal in deep bedrock

[3] Assume currently impacted volume of aquifer cannot be completely rehabilitated (TI)

[4] Assume protection of off-site residents is a clear + required objective

[5] Assume protection of off-site and on-site workers is clear + required objective

(impact of contaminants on surface water users), 6A (impact of vapors on off-site residents), and 7A (impact of vapors on workers), no action is considered to be an unacceptable remedial option. The results of the risk assessment dictate that these SPR linkages must be broken in some way.

13.4.1.2 Constraints Analysis

Constraints to remediation can take many different forms: time pressures, physical limitations on placement of equipment, technologicallimitations, land ownership, and property access constraints are just some of the issues that could affect the selection of a remedial objective. Also, some objectives, such as removal of all contamination from the aquifer, may be technologically impractical/infeasible (TI).

Table 13.5 identifies the chief constraints for each risk linkage in this example. It is assumed that there are no time constraints on remedial activity. The property owner wishes to redevelop and sell the site, but this is not expected to result in demands for immediate sale. Any anticipated groundwater remedial activity can take place over an extended period, without hampering site development. There are, however, significant physical constraints to the remediation planning. Remedial activity is confined to the site itself, and no remedial activity can take place on adjacent properties, because of access and legal restrictions.

Technological constraints have been identified at the site. Based on the detailed site investigation work completed, including extensive coring, sampling, and geophysical logging of the Sherwood sandstone, removal of any significant proportion of the NAPL from the fractured bedrock is considered to be technologically infeasible. Given the current techniques and tools available and the current level of knowledge about the remediation of NAPL in fractured bedrock, it is presently unrealistic to set an objective of complete source removal in linkages involving this source. In Table 13.5, these constraints have been used to justify listing source elimination options as "unacceptable objectives."

13.4.1.3 Objective Short-List Development

After the user has proceeded through the preceding substeps, a short-list of generally acceptable and viable remedial or risk management objectives emerges. As shown in Table 13.6, a short-list of remedial objectives now exists for each risk linkage. Objectives that are clearly inappropriate or that would not be acceptable to regulators have been eliminated. Constraints have been applied.

13.4.2 STEP A2 — DETERMINE APPROACH SHORT-LIST FOR EACH OBJECTIVE

Next, likely remedial approaches able to reach each of the remedial objectives shortlisted in Table 13.6 were developed, considering site conditions and the results of the risk assessment. In this example, 20 different approaches (alone or in combination) were evaluated on a preliminary basis, considering the types of technologies that would be entailed and their overall cost and practicality at the site.

TABLE 13.6
Remedial Objective Options Short-List

	Receptor	Pathway	Source	Remedial Objective Options
1A	River	NAPL via superficial deposits	NAPL in superficial deposits	C, D, E
1B	River	Aqueous phase via superficial deposits	NAPL in superficial deposits	B, C, D, E
2	River	Aqueous phase via bedrock	NAPL in bedrock	B, C
3A	Surface water user	NAPL via superficial deposits	NAPL in superficial deposits	A, D, E
3B	Surface water user	Aqueous phase via superficial deposits	NAPL in superficial deposits	A, D, E
4	Surface water user	Aqueous phase via bedrock	NAPL in bedrock	A, E
5	Aquifer and users	Aqueous phase via bedrock	NAPL in bedrock	A, B, E, F
6A	Residents off-site	Vapor from NAPL in superficial deposits	NAPL in superficial deposits	A
6B	Residents off-site	Vapor from NAPL in bedrock	NAPL in bedrock	A
7A	Workers on and off-site	Vapor from NAPL in superficial deposits	NAPL in superficial deposits	A
7B	Workers on and off-site	Vapor from NAPL in bedrock	NAPL in bedrock	A

13.4.2.1 Available Remedial Approaches

As discussed earlier, remedial approaches focus on how the SPR risk linkage is to be broken to achieve the specified objective. Benefits could vary, depending on the remedial approach selected. Thus, the benefits part of the economic analysis requires that remedial approaches be examined. Approach selection will also have a direct impact on costs, because different approaches will typically involve different technologies. Determining the best approach means that alternative approaches must be considered for each objective and the cost and benefits of each evaluated at a high level. Only then can the most economic objective for a given SPR linkage be determined.

A list of available remedial approaches was compiled and is presented in Table 13.7. Approaches are categorized as applying to source removal, pathway elimination, or receptor isolation.

13.4.2.2 Develop Short-List of Remedial Approaches

If one particular remedial approach is the clear choice for achieving the objective, due either to the particular physical or environmental considerations at the site or

TABLE 13.7
Available Remedial Approaches List

Source approaches	S1: Remove NAPL in surficials
	S1P: Partially remove NAPL in surficials
	S2: Remove NAPL in bedrock
	S2P: Partially remove NAPL in bedrock
Pathway approaches	P1D: Contain dissolved phase in surficials
	P1N: Contain NAPL in surficials
	P2D: Contain dissolved phase in bedrock
	P2N: Contain NAPL in bedrock
	P4N: Contain vapor at limit of NAPL plume
	P4D: Contain vapor at limit of dissolved-phase plume
Receptor approaches	R1: Collect and treat discharge to river at river
	R2T: Treat water at point of collection (surface water user)
	R2R: Provide surface water user with another source
	R3T: Treat groundwater at point of abstraction
	R3R: Provide replacement source of water for lost groundwater
	R4M: Indoor air monitoring in houses
	R4P: Install vapor protection systems in affected houses
	R4E: Purchase affected houses at full market value
	R5P: Develop protective working practices for workers in affected areas
Institutional management	N3MNA: Allow plume to stabilize or remediate with natural attention

to specific legal or regulatory requirements for action, it is selected. As shown in Table 13.8, clear choices of remedial approach exist in several cases. For linkage 1A (NAPL reaching the river through the superficial materials or superficial deposits), the remedial objectives of removing all (D) or part (E) of the contamination can best be satisfied by removing all or part of the NAPL contained in the superficial deposits (approaches S1 and S1P), respectively.

Certain remedial approaches are also not acceptable, either to one or to many stakeholders. These can be ruled out from further consideration. In the example (Table 13.8), several approaches are eliminated. For linkage 1A (NAPL reaching the river through the superficial materials or superficial deposits), the remedial objectives of eliminating all impact to the current receptor (C) through collection and treatment of all discharge to the river (through the use of a cofferdam structure, for instance) is not practical, because it violates the constraint about working off site. Note that in Table 13.8, individual approaches and combinations of approaches are listed. The result of screening out clearly unacceptable or clearly mandated approaches is a rather lengthy list of approaches and approach combinations. Note that several remedial approaches, such as S1P (limited excavation of contaminated hotspots in surficial sediments) and N3MNA (institutional management through natural attenuation and monitoring of dissolved-phase plume) apply to many of the risk linkages.

TABLE 13.8
Remedial Approach Analysis Matrix

SPR	Remedial Objective	Step A2-1	Step A2-2	Step A2-3	Step A2-5
Linkage	Option	Possible Approach	Clear Approach	Inappropriate Approach	Acceptable Approaches Before Constraints Applied
1A	C	S1, S1P, P1N, R1	—	R1	S1, S1P, P1N
	D	S1	S1	—	S1
	E	S1P	S1P	—	S1P
B	B	S1, S1P, P1D, R1	—	R1	S1, S1P, P1D
	C	S1, P1D, R1	—	R1	S1+P1D, P1D
	D	S1, S2	S1	S2	S1
	E	S1, S1P, S2, S2P	—	S2	S1, S1P, S2P
2	B	S2, S2P, P2D, P2N, R1	—	R1, S2	S2P, P2D, P2N
	C	S2, S2P, P2D, P2N, R1	S2P+P2N+S2D	R1, S2	S2P+P2D+P2N
3A	A	S1, S1P, P1N, R1, R2T, R2R	—	R1, R2R	S1, S1P, P1N, R2T
	D	S1, S2	S1	S2	S1
	E	S1P	S1P	—	S1P
3B	A	S1, S1P, P1N, P1D, R1, R2T, R2R	—	R1, R2R	S1, S1P, P1N+P1D, R2T
	D	S1, S2	S1	S2	S1
	E	S1P, S2P, S1, S2	—	S2	S1, S1P, S2P
4	A	S2, S2P, P2D+P2N, R1	—	R1, S2, R2R	S2P, P2D+P2N, R2T
	E	S2P	S2P	—	S2P

TABLE 13.8
Remedial Approach Analysis Matrix (continued)

SPR	Remedial Objective	Step A2-1	Step A2-2	Step A2-3	Step A2-5
Linkage	Option	Possible Approach	Clear Approach	Inappropriate Approach	Acceptable Approaches Before Constraints Applied
5	A	S2,S2P,P2D+P2N,R3T,R3R	—	S2	S2P, P2D+P2N, R3T, R3R
	B	S2, S2P, P2D, P2N	—	S2	S2, S2P, P2D, P2N
	E	S2P	S2P	—	S2P
	F	N3MNA	N3MNA	—	N3MNA
6A	A	S1, S1P, P1N, R4E, P4M+R4P	—	—	S1, S1P, P1N, P4N, R4M+R4P, R4E
	D	S1, P4N	—	—	S1, P4N
	E	N3MNA, R4M	—	—	N3MNA, R4M
6B	A	S2, S2P, P2N, P4N, R4M+R4P, R4E	—	S2	S2P, P2N, P4N, R4M+R4D, R4E
	E	N3MNA, R4M	—	—	N3MNA, R4M
7A	A	S1, S1P, P1N, R4E, P4M+R4P, R5P	R5P	—	S1, S1P, P1N, P4N, R4M+R4P, R4E, R5P
	D	S1, P4N, R5P	R5P	—	R5P
	E	N3MNA, R4M	—	—	N3MNA, R4M
7B	A	S2, S2P, P2N, P4N, R4M+R4P, R4E, R5P	R5P	S2	R5P
	E	N3MNA, R4M	—	—	N3MNA, R4M

However, as discussed regarding remedial objectives, there will always exist constraints that will limit how an approach can be implemented, when, and where. These constraints should be identified, based on site-specific conditions. In this example, the constraints are the same as those listed for the remedial objective analysis and are summarized in Table 13.7. By applying these constraints, the list of remedial approaches that can reasonably be applied for each risk linkage is further reduced.

Consideration of relative costs, at this point in the analysis, provides a precursor to step B2 (development of high-level cost functions for each approach). Table 13.9 provides a more detailed evaluation of which approaches are available for reaching each objective, for each SPR linkage. By using relative cost, the short-list of applicable approaches for further analysis can be further reduced, helping to simplify the analysis.

Note that in the case of vapor impacts on workers on and off site, approach R5P (development of safe working practices for workers potentially exposed to vapor) would be by far the lowest-cost approach and would lead to selection of remedial objective A for those risks. In addition, the results of the risk assessment have indicated that the risk of vapor exposure to workers on site is low.

In some cases, such as when the relative costs are easily identified as markedly different among approaches, the user may be able to move directly to selecting the least-cost approach. This would eliminate the need to conduct the more detailed analysis in step B3 (determining the economically optimal approach for each objective), and the user could move directly to step C (CBA). Step B3 is used only when different remedial approaches will clearly entail not only different costs, but also different *benefits*.

In all, after looking at relative costs and applying constraints, only eight remedial approaches (alone or in combination) have been found worthy of more detailed analysis by CBA for this site. The short list of remedial approaches is provided in Table 13.10. In practice, getting to this stage of a remedial analysis is often done quickly, based on the knowledge and experience of the practitioner. What has been illustrated here is simply a formalized system for reaching the same point. The authors also acknowledge that many other methodologies are available for developing similar short-lists; in practice, these may be substituted for the methodology described in Part III and illustrated here. However, the key element in the analysis presented is the relationship among risks, the remedial objectives meant to manage those risks, and the remedial approach options that can be used to achieve the objectives.

13.4.2.3 Relationships among Multiple Objectives

In many situations, there will exist several remedial objectives, each of which will be a desired outcome. This is clearly the case in the example site, where 11 SPR linkages have been identified. Various approach options for achieving each objective will have different impacts on each objective. The relationships can be described as falling into the categories presented in Table 13.11 and discussed in Chapter 12.

TABLE 13.9
Shortlist of Applicable Remedial Approaches

Risk Linkage

		Relative Cost	1A			1B				2		3A			3B			4		5				6A		6B			7A			7B		#
	Remedial Approach		C	D	E	B	C	D	E	B	C	A	D	E	A	D	E	A	E	A	B	E	F	A	E	A	D	E	A	D	E	A	E	#
S1	Remove NAPL in surficials	H	✓	✓		✓	✓	✓	✓			✓	✓	✓	✓	✓	✓											✓	✓						14
S1P	Partially remove NAPL in surficials	M	✓	✓				✓				✓	✓	✓		✓	✓										✓			✓					10
S2	Remove NAPL in Bedrock	VH																			✓														1
S2P	Partially remove NAPL in bedrock	M								✓	✓					✓	✓	✓	✓	✓		✓	✓		✓										10
S1+P1D	Remove NAPL in surficials + Partially remove NAPL in surficials	VH					✓								✓																				2
S2P+P2D +P2N	Partially remove NAPL in bedrock + Contain dissolved phase in bedrock + Contain NAPL in bedrock	H									✓																								1
P1D	Contain dissolved phase in surficials	H	✓			✓	✓																												3
P1N	Contain NAPL in surficials	H										✓														✓			✓						3
P2D	Contain dissolved phase in bedrock	H								✓											✓														2
P2N	Contain NAPL in bedrock	H									✓										✓			✓								✓		4
P4N	Contain vapor at limit of NAPL plume	M																						✓		✓	✓		✓	✓		✓		6
P4D	Contain vapor at limit of dissolved-phase plume	M																																0

Code	Description	Rating	Number
P2D+P2N	Contain dissolved phase in bedrock + Contain NAPL in bedrock		2
R1	Collect and treat discharge to river at river	VH	0
R2T	Treat water at point of collection (surface water user)	H	3
R2R	Provide surface water user with another source	H	0
R3T	Treat groundwater at point of abstraction	H	0
R3R	Provide replacement source of water for lost groundwater	VH	1
R4M	Indoor air monitoring in houses	L	1
R4P	Install vapor protection systems in affected houses	H	0
R4E	Purchase affected houses at full market value	H	4
R5P	Develop protective working practices for workers in affected areas	L	3
P1N+P1D	Contain NAPL in surficials + Contain dissolved phase in surficials		0
R4M+R4P	Indoor air monitoring in houses + Install vapor protection systems in affected houses	LTH	7
N3 MNA	Abandon damaged of + MNA	L–M	5

Notes: V means the remedial approach can manage the risk; number means the total number of risks managed by that remedial approach.

TABLE 13.10
Short-Listed Remedial Approach Alternatives

Designation	Approach Description	Relative Cost
	Source Methods	
S1	Remove NAPL in superficial deposits	High
S1P	Partially remove NAPL in superficial deposits	Moderate
S2P	Partially remove NAPL in bedrock	Moderate
	Pathway Methods	
P1D	Contain dissolved phase in superficial deposits	High
P2D	Contain dissolved phase in bedrock	High
P4N	Contain vapor at limit of NAPL plume	Moderate
N3MNA	Allow natural attenuation to act on plume	Low to moderate
	Receptor Methods	
R5P	Implement protective working practices for workers to reduce vapor risk	Low

TABLE 13.11
Multiple Objective Relationships

1. *No effect (0)*: The approach under consideration has no effect on another objective or another approach.
2. *Positive effect(+)*: The approach will have a positive effect on the implementation of another approach or objective.
3. *Contingent (C)*: Some approaches will be contingent on successful implementation of another approach. For instance, relying on natural attenuation processes to deal with a secondary objective may require that the subsurface source (LNAPL on groundwater in the example) be removed first.
4. *Achieves the other goal (A)*: The approach being considered for one objective actually accomplishes another goal at the same time.
5. *Negative effect (-)*: The approach has a net negative effective on another objective or approach.

When multiple objectives are being considered, as in this case, a matrix format can be used to present the complex interactions between objectives and approaches. For this example, the objective–approach relationship matrix is shown in Table 13.12. Although the matrix appears complex, the results can be quickly tabulated to provide a picture of which approaches are most beneficial as a whole, contributing positively to the management of the greatest number of SPR linkages. For example, remedial approach S1P (partial removal of NAPL from the superficial deposits) applied to

TABLE 13.12
The Matrix of Objective–Approach Relationships

		SPR Linkage										
SPR Linkage	Approach/ Objective	1A	1B	2	3A	3B	4	5	6A	6B	7A	7B
1A	SIP-E,A	X	+	+	A	+	+	+	A	+	A	+
1B	SIP-B,E	+	X	+	A	+	+	+	A	+	A	+
	PID-C	A	X	0	A	A	0	0	A	0	0	0
2	S2P-B	0	0	X	0	0	+	+	0	+	0	+
3A	SIP-A,E	+	+	+	X	+	+	+	A	+	A	+
3B	SIP-A,E	+	+	+	A	X	+	+	A	+	A	+
4	SIP-A,E	+	+	+	A	+	X	+	A	+	A	+
5	S2P-A,B,E	0	0	A	0	0	+	X	0	+	0	+
	N3MNA-F	0	0	0	0	0	0	X	0	0	0	0
6A	S1P-A	+	A	+	A	+	+	+	X	+	A	+
	P4N-A	0	0	0	0	0	0	0	X	A	A	A
	N3MNA-E	0	0	0	0	0	+	A	X	0	0	0
6B	S2P-A	0	0	A	0	0	+	+	0	X	0	+
	P4N-D	0	0	0	0	0	0	0	A	X	A	+
7A	R5P-A	0	0	0	0	0	0	0	A	+	X	+
	R5P-D	0	0	0	0	0	0	0	0	0	X	+
	N3MNA-E	0	0	0	0	0	+	A	A	0	X	0
7B	R5P-A	0	0	0	0	0	0	0	0	0	0	X
	P4N-A	0	0	0	0	0	0	0	A	A	A	X
	N3MNA-E	0	0	0	0	0	+	A	A	0	A	X

Note: Refer to Table 13.11 for symbol explanations (A, 0, +, –).

SPR linkage 1A (NAPL to the river) contributes positively to the management of 7 of the other 11 SPR linkages and achieves the remedial objectives outright for three of the others. This is an important consideration when looking at complex sites of this nature. In contrast, applying monitored natural attenuation (N3MNA) to achieve an institutional management objective for SPR linkage 5 (dissolved-phase contamination impacting the aquifer) has no effect whatsoever on any of the other SPR linkages.

Considering the relationships among the various approaches described previously and a preliminary high-level consideration of the relative costs of the various approaches, a remedial approach is selected for each of the remedial objectives shortlisted for each SPR linkage. These are shown in Table 13.13. For SPR linkage 1A, for example, both remedial objectives D and E are best reached using approach S1P,

TABLE 13.13
Remedial Approach Short-List Matrix

SPR Linkage	Remedial Objective Options	S1	S1P	S2P	P1D	P2D	P4N	R5P	N3MNA
1A	C				X				
	D		X						
	E		X						
1B	B		X						
	C		X						
	D	X							
	E		X						
2	B					X			
	C					X			
3A	A		X						
	D	X							
	E		X						
3B	A		X						
	D	X							
	E		X						
4	A		X						
	E			X					
5	A			X					
	B			X					
	E			X					
	F								X
6A	A		X						
6B	A			X					
7A	A							X	
7B	A							X	

Note: The most appropriate approach for each objective option is indicated by an X.

and objective C is best reached using approach P1D. Selection of the most appropriate approach for each objective at this stage means that step B3 (determining the economically optimal approach for each objective) is skipped. This may occur, at the user's discretion, when it is clear that the various remedial approaches being considered will not result in the realization of significantly different benefits — benefits vary depending on which remedial objective is achieved, not how it is achieved. In most situations, this is likely to be the case, significantly streamlining and simplifying the analysis.

13.4.3 Step B1 — High-Level Benefits of Each Objective/Approach

A short-list of acceptable remedial objectives for each SPR linkage has been developed in step A. Now, the economic benefits of each objective may be determined, en route to a partial CBA (steps B3 and C), allowing an economic comparison to be made. This is where the economic valuation techniques discussed in Part I are used to monetize the results of applying the various short-listed remedial approaches.

13.4.3.1 Link with Risk Assessment

The results of the risk assessment were used in step A1 to assist in selection and short-listing of remedial objectives. At this stage, risk assessment is used to provide specific information about identified receptors, the degree of current and anticipated impact, and the relative importance placed on those receptors (Table 13.14). This information is used to help quantify the impacts on those receptors, which in turn allows those impacts to be monetized, using the techniques discussed in Part I.

Based on this analysis, the priority SPR linkages are considered to be 6A and 7A (vapors impacting workers and residents off site, generated by NAPL in superficial deposits) and 5 (dissolved-phase contaminant migration in the groundwater aquifer).

13.4.3.2 Possible Benefits of Each Objective/Approach

Using the information from the risk assessment from substep B1-1, possible benefits are identified that may accrue from successful realization of the objectives (using the particular approach). Possible financial (private) benefits are listed in Table 13.15A. Possible external benefits (other than the problem holder's) are listed in Table 13.15B. Benefits are classified as very likely (VL), likely (L), or not likely (NL) to occur.

In both tables, we only consider those benefits that are very likely (VL) to be realized. In addition, as explained earlier, quantification of health benefits is beyond the scope of this analysis (because all remedies are considered to be protective of human health as a minimum), so such benefits are not considered further in the monetary analysis.

What is apparent from these two tables is that the analysis quickly reduces to a few key benefits that are likely to be realized:

- *Increased site property value* — The site can be sold and redeveloped, earning an income for the current property owner (readily monetized).
- *Increased property value in the immediate neighborhood of the site* — Remediation and redevelopment of the site will eliminate blight from the area and will likely push local property values higher (readily monetized).
- *Reductions in corporate liability* and the possibility of legal action against the site owner and increased public relations value to the site owner —

TABLE 13.14
Receptor Risk Information for Benefits Assessment

	Receptor	Pathway	Contaminant Impacting	Expected Impact
1A	River	NAPL via superficial deposits	Pure-phase NAPL	NAPL sheen on river, some possible odor. Benzene mass flux up to 50 g/day. Expected concentration above EQS (freshwater) after dilution. Impact limited to site area reach of river.
1B	River	Aqueous phase via superficial deposits	Dissolved benzene and naphthalene	Benzene mass flux 3.5 g/day. Expected benzene concentration in river 0.35 μg/l (below EQS). Diluted in river.
2	River	Aqueous phase via bedrock	Dissolved benzene and naphthalene	Benzene flux < 3.5 g/day. Expected benzene concentration in river 0.35 μg/l (below U.K. freshwater standard). Diluted in river.
3A	Surface water user	NAPL via superficial deposits	Dissolved benzene and naphthalene	Expected benzene concentration in river 0.35 μg/l (below U.K. freshwater standard). Diluted in river.
3B	Surface water user	Aqueous phase via superficial deposits	Dissolved benzene and naphthalene	Expected benzene concentration in river 0.35 μg/l (below U.K. freshwater standard). Diluted in river.
4	Surface water user	Aqueous phase via bedrock	Dissolved benzene and naphthalene	Benzene flux < 3.5 g/day. Expected benzene concentration in river 0.35 μg/l (below U.K. freshwater standard). Diluted in river.
5	Aquifer and users	Aqueous phase via bedrock	Dissolved benzene and naphthalene	Benzene at PWS borehole 500 m east of site, predicted at 100 mg/l, considerably above E.U. drinking-water limits. Presence of plume eliminates potential abstraction of 4600 m³/day.
6A	Residents off site	Vapor from NAPL in superficial deposits	Benzene and naphthalene vapors	Vapor concentrations in open excavations and basements may exceed OHS exposure limits.
6B	Residents off site	Vapor from NAPL in bedrock	Benzene and naphthalene vapors	Vapor concentrations unlikely to contribute to risk at surface.
7A	Workers on and off site	Vapor from NAPL in superficial deposits	Benzene and naphthalene vapors	Vapor concentrations in open excavations and basements may exceed OHS exposure limits.
7B	Workers on and off site	Vapor from NAPL in bedrock	Benzene and naphthalene vapors	Vapor concentrations unlikely to contribute to risk at surface.

TABLE 13.15A
Possible Financial (Private) Benefits

SPR	Receptor	Remedial Objective Options	Possible Benefits — Financial Analysis	Likelihood of Realizing Benefit
1A	River	C	Reduce corporate liability	L
			Reduce possibility of litigation	NL
			Positive public relations value	NL
		D	**Reduce corporate liability**	**VL**
			Reduce possibility of litigation	L
			Positive public relations value	L
			Increase property value	**VL**
		E	Reduce corporate liability	L
			Reduce possibility of litigation	L
			Positive public relations value	L
			Increase property value	**VL**
1B	River	B	Reduce corporate liability	NL
			Reduce possibility of litigation	NL
			Positive public relations value	NL
		C	Reduce corporate liability	L
			Reduce possibility of litigation	NL
			Positive public relations value	NL
		D	**Reduce corporate liability**	**VL**
			Reduce possibility of litigation	L
			Positive public relations value	L
			Increase property value	**VL**
		E	Reduce corporate liability	L
			Reduce possibility of litigation	L
			Positive public relations value	L
			Increase property value	**VL**
2	River	B	Reduce corporate liability	NL
			Reduce possibility of litigation	NL
			Positive public relations value	NL
		C	Reduce corporate liability	L
			Reduce possibility of litigation	NL
			Positive public relations value	NL
3A	Surface water users	A	Reduce corporate liability	NL
			Reduce possibility of litigation	NL
			Positive public relations value	NL
		D	**Reduce corporate liability**	**VL**
			Reduce possibility of litigation	L
			Positive public relations value	L
			Increase property value	**VL**
		E	Reduce corporate liability	L
			Reduce possibility of litigation	L
			Positive public relations value	L
			Increase property value	**VL**

TABLE 13.15A
Possible Financial (Private) Benefits (continued)

SPR	Receptor	Remedial Objective Options	Possible Benefits — Financial Analysis	Likelihood of Realizing Benefit
3B	Surface water users	A	Reduce corporate liability	NL
			Reduce possibility of litigation	NL
			Positive public relations value	NL
		D	**Reduce corporate liability**	**VL**
			Reduce possibility of litigation	L
			Positive public relations value	L
			Increase property value	**VL**
		E	Reduce corporate liability	L
			Reduce possibility of litigation	L
			Positive public relations value	L
			Increase property value	**VL**
4	Surface water users	A	Reduce corporate liability	NL
			Reduce possibility of litigation	NL
			Positive public relations value	NL
		E	Reduce corporate liability	NL
			Reduce possibility of litigation	NL
			Positive public relations value	NL
			Increase property value	NL
5	Aquifer	A	Reduce corporate liability	NL
			Reduce possibility of litigation	NL
			Positive public relations value	L
		B	Reduce corporate liability	NL
			Reduce possibility of litigation	NL
		E	Reduce corporate liability	L
			Reduce possibility of litigation	L
			Positive public relations value	NL
			Increase property value	NL
		F	No benefits	
6A	Residents	A	**Reduce corporate liability**	**VL**
			Reduce possibility of litigation	**VL**
			Positive public relations value	**VL**
			Increase property value	**VL**
6B	Residents	A	Reduce corporate liability	NL
			Reduce possibility of litigation	NL
			Positive public relations value	L
			Increase property value	NL
7A	Workers	A	**Reduce corporate liability**	**VL**
			Reduce possibility of litigation	**VL**
			Positive public relations value	**VL**
			Increase property value	**VL**
7B	Workers	A	Reduce corporate liability	NL
			Reduce possibility of litigation	NL
			Positive public relations value	L
			Increase property value	NL

TABLE 13.15B
Possible External Benefits

SPR	Receptor	Remedial Objective Options	Possible Benefits — Economic Analysis	Likelihood of Realizing Benefit
1A	River	C	**Avoid ecological damage to river**	**VL**
			Improve recreational value of river	NL
			Increased property value in neighborhood	NL
		D	Avoid ecological damage to river	NL
			Improve recreational value of river	NL
			Increased property value in neighborhood	**VL**
			Option value — redevelopment of brownfield	L
		E	Avoid ecological damage to river	NL
			Improve recreational value of river	NL
			Increased property value in neighborhood	**VL**
			Option value — redevelopment of brownfield	L
1B	River	B	Avoid ecological damage to river	VL
			Improve recreational value of river	NL
			Increased property value in neighborhood	NL
		C	Avoid ecological damage to river	NL
			Improve recreational value of river	NL
			Increased property value in neighborhood	NL
		D	Avoid ecological damage to river	NL
			Improve recreational value of river	NL
			Increased property value in neighborhood	NL
			Option value — redevelopment of brownfield	NL
		E	Avoid ecological damage to river	NL
			Improve recreational value of river	NL
			Increased property value in neighborhood	NL
			Option value — redevelopment of brownfield	NL
2	River	B	**Avoid ecological damage to river**	**VL**
			Improve recreational value of river	NL
			Increased property value in neighborhood	NL
		C	Avoid ecological damage to river	NL
			Improve recreational value of river	NL
			Increased property value in neighborhood	NL
3A	Surface water users	A	Improve recreational value of river	NL
			Health benefits to river users	NL
		D	Avoid ecological damage to river	NL
			Improve recreational value of river	NL
			Increased property value in neighborhood	**VL**
			Option value — redevelopment of brownfield	L

TABLE 13.15B
Possible External Benefits (continued)

SPR	Receptor	Remedial Objective Options	Possible Benefits — Economic Analysis	Likelihood of Realizing Benefit
		E	Avoid ecological damage to river	NL
			Improve recreational value of river	NL
			Increased property value in neighborhood	**VL**
			Option value — redevelopment of brownfield	L
3B	Surface water users	A	Improve recreational value of river	NL
			Health benefits to river users	NL
		D	Avoid ecological damage to river	NL
			Improve recreational value of river	NL
			Increased property value in neighborhood	**VL**
			Option value — redevelopment of brownfield	L
		E	Avoid ecological damage to river	NL
			Improve recreational value of river	NL
			Increased property value in neighborhood	**VL**
			Option value — redevelopment of brownfield	L
4	Surface water users	A	Improve recreational value of river	NL
			Health benefits to river users	NL
		E	Avoid ecological damage to river	NL
			Improve recreational value of river	NL
			Increased property value in neighborhood	NL
			Option value — redevelopment of brownfield	NL
5	Aquifer	A	Health benefits to water users	NL
			Financial benefits to water company	NL
		B	Health benefits to water users	NL
			Financial benefits to water company	NL
		E	Health benefits to water users	NL
			Reduce aquifer damage, drop in aquifer TEV (total economic value)	**VL**
		F	Health benefits to water users	NL
			Reduce aquifer damage, drop in aquifer TEV	**VL**
6A	Residents	A	Health benefits to residents	L
			Increased property value in neighborhood	**VL**
6B	Residents	A	Health benefits to residents	NL
			Increased property value in neighborhood	NL
7A	Workers	A	Health benefits to residents	L
			Increased property value in neighborhood	**VL**
7B	Workers	A	Health benefits to residents	NL
			Increased property value in neighborhood	NL

These benefits can be seen as having value to the property owner but cannot strictly be counted within the economic analysis, because these are likely to be transfer payments (transfers from one party to another with no net change in society's welfare).

- *Option value on brownfield redevelopment* — If the contaminated site is redeveloped, a green-field site elsewhere is retained in its natural state. Recent studies have suggested that the public is willing to pay for preventing the development of green-field sites by reusing "brown land."[4] This benefit is currently difficult to monetize, however.
- *Reduction in aquifer damage* and the resulting protection of the aquifer's total economic value (TEV) — By preventing further degradation of the aquifer over time or by remediating over time the volume of aquifer already impacted, an increase in the aquifer's TEV is realized (partially monetizable).
- *Prevention of further damage to the river* — Manifested as a gradual improvement in river quality designation over time. However, because the river is already of very poor quality and located within a heavily industrialized area, efforts to reduce the relatively small amounts of contaminant entering from the site are unlikely to realize more than relatively small benefits (partially monetizable).

13.4.3.3   Identify and Assess Quantifiable Benefits

Of the likely benefits identified previously, only a few are readily quantified in monetary terms. The analysis indicates that there are only four main readily monetized benefit categories, if the damage-avoided concept and economic analysis are followed. In this case, benefits of remediation accrue:

- If damage to the river and its users is avoided
- If damage to the aquifer and its users is avoided
- If damage to the property itself is avoided
- If damage to the neighboring properties is avoided

The methodology does not incorporate directly the benefits of avoiding damage to human health, as discussed, so these are not directly monetized. Instead, benefits of remediation to workers and off-site residents are valued by the increase in property value realized by remediation or by the elimination of negative effects on property values that accrue due to remediation of the site. The benefits of remediation that could be readily monetized for this example are listed in Table 13.16. In all cases, these are the *additional benefits* that accrue as a result of remediation.

TABLE 13.16
Remedial Benefits Summary Table

Benefit Category	One-Time Benefit (in present value)	Annual Damage Avoided (Per Year)	Planning Horizon (Yrs)	Valuation Method
Prevention of river water quality degrading to the point where the next lowest U.K. river category is reached	N/A	Negligible < $0.018 million (£0.01 million)	20	Willingness to pay to avoid further decline of river water quality (estimate taken from literature).
Lost potential water production from portion of aquifer rendered unusable by contamination in bedrock	N/A	$0.18 (£0.1)/m³ of abstraction; for 4600 m³/day lost production	20 for restoration of plume through natural attenuation once source is removed	Modeled lost potential water production due to presence of plume, multiplied by current commercial market value of water in U.K. (actual market value as a lower bound estimate). Assume rate for aquifer amelioration once source removal efforts in place.
Sale of property as commercial site, once made suitable for such use	$1.89 million (£1.05 million)	N/A	One-time benefit	Current market value of commercial property in this part of the U.K.
Recovery of property value in sites within the vicinity of the site, as a result of remediation	$0.72 million (£0.40 million)	N/A	One-time benefit	Notional improvement of 10% in average property value in adjacent sites, based on current market values, due to elimination of blight.

Note: N/A = not applicable

13.4.3.4 Develop Threshold Benefit–Time Function

The readily quantifiable benefits listed previously will exhibit time-variant behavior. The benefits for the property and neighboring property values are considered one-time payments. However, river and aquifer quality benefits would accrue over time. In the case of the aquifer, modeling has shown that the plume appears to be stable, so no additional damage is expected to accrue over time, even without remediation.

TABLE 13.17
Twenty-Year Cumulative Remedial Benefits at 5% Discount Rate

Economic Analysis Scenarios	PV Benefit ($M)
Full aquifer restoration, immediate start	4.34
50% aquifer restoration, immediate start	3.59
25% aquifer restoration, immediate start	3.22
25% aquifer restoration over 20 years, start remediation in 5 years	2.29
25% aquifer restoration over 20 years, start remediation in 10 years	1.71

Thus, remedial activities that serve to improve aquifer quality over time or to reduce damage to the aquifer are considered benefits over time. Several possible scenarios are illustrated, considering the lack of available data with which to predict the actual positive impact of remediation on the aquifer. It is assumed that remedial activities that reduce mass flux to the existing dissolved-phase plume will result in improvements to aquifer water quality and that these will be realized incrementally over time in a linear fashion. Thus, for the base case with 25% plume amelioration, linearly over 20 years (1.25% improvement per year for 20 years), at a 5% discount rate, the total benefits over 20 years would be $0.37 million (£0.21 million). For complete aquifer restoration over 20 years (unlikely), total benefits would be $1.48 million (£0.83 million).

Total present value of all accumulated benefits fully realized over 20 years, assuming a 5% discount rate, starting the remediation in year 1, and assuming complete aquifer restoration, is approximately $4.34 million (£2.41 million). Table 13.17 provides total cumulative remedial benefits over the 20-year planning horizon for a variety of aquifer restoration rates and a variety of remedial start times. All scenarios assume that property value increases are realized and the meager river-quality improvement benefits are realized. This information shows how varying assumptions regarding aquifer amelioration can affect the total realized benefits and also reveals the impact on benefits of delaying the start of remediation. At a 5% discount rate, delaying the start of work by ten years reduces the benefit of the remediation by almost half (from $3.22 million to $1.71 million for the 25% aquifer amelioration case).

13.4.3.5   Identify and Assess Nonquantifiable Benefits

As discussed in Part II, it is also possible to identify benefits that cannot be readily monetized. Many such benefits are considered important and, if properly quantified, of significant value. Such benefits need to be included in the overall analysis. Benefits are subjectively assessed based on whether they are expected to increase overall benefits in general, by a negligible (0), marginal (+), substantial (++), or considerable (+++) amount. This can be accomplished, in part, by gauging the expected contribution of the nonquantifiable benefits against the magnitude of overall quantifiable

TABLE 13.18
Nonquantifiable Benefits

Benefit	Private/ External	Likelihood	Relative Benefit Impact	Applicable to Approach	Uncertainty	Comments
Option value on green- field site realized	E	L	+	S1, S1P	Medium	Redevelopment of the site means that a green-field site is preserved. Research has demonstrated WTP for this option.
Public relations value to problem holder	P	L	+	S1, S1P, S2P, P1D, P2D, P4N, R1, N3MNA	Medium	Very difficult to quantify the positive effect on the problem holder's business.
Reduced liability to problem holder	P	VL	++	S1, S1P, S2P, P2D, P4N,	Low	Liability for the site largely eliminated once sold. Provision (if present) can be removed from firm's balance sheet. Not captured by increase in property value.

benefits. Benefits classified as likely (L) or very likely (VL) in Tables 13.15A and 13.15B are considered. Table 13.18 indicates their relative impact as judged by the authors. As discussed in Part III, the assessment of the importance of such benefits and their ranking is highly subjective and must be considered carefully and with the benefit of considered expertise.

13.4.3.6 Apply Nonquantified Benefits

In simple terms, the nonquantifiable benefits mentioned previously are considered to have a reasonable likelihood of substantially increasing the benefits of remediation. In particular, in this example, the reduction in the problem holder's liability, as a result of successful remediation and sale of the property to another party, is expected to be substantial. As discussed in Part III, *substantial* would mean on the order of the monetized benefits. Thus, we might expect that the actual total economic benefit of the remediation is perhaps up to a maximum of twice the monetized "threshold" benefit.

13.4.4 Step B2 — Assess High-Level Cost Function for Each Approach

13.4.4.1 Select Likely Least-Cost Technology to Achieve Approach

This step provides the conceptual link between the objective/approach level of analysis and the approach/technology level. The user must develop preliminary costs for implementing each approach being considered in the analysis. This has already begun in Table 13.9, with qualification of the relative costs of each approach. Now, a preliminary estimate of least-cost implementation for each approach is required. At this stage, the analysis of costs does not have to be overly detailed, especially considering that benefit analysis (to which these costs will be compared later) is unlikely to be complete. Table 13.19 provides preliminary cost estimates for the least-cost implementation of the short-listed approaches.

13.4.4.2 Qualify Probability of Success of Each Approach

A probability of success is assigned to each approach using the selected least-cost technology. The qualifying system for remedial technologies described in the framework (Part III) is used. Definitions of the designation are provided in Table 13.20, and how these are applied in this case is shown in Table 13.21.

13.4.5 Step B3 — Determine Economically Optimal Approach for Each Objective

As discussed earlier, the selection of the most appropriate approach for each of the short-listed objectives for each SPR linkage was made in step A2 (see Table 13.13). In all but the most complex and involved analyses, this is expected to be the case. Most often, selection of the most appropriate approach for each objective can be made based on available relative cost information and the results of steps A1 and A2.

Table 13.13 shows which approaches are considered most appropriate for each objective. Interestingly, Table 13.13 shows that a few approaches are chosen for the majority of the short-listed objectives. Approaches S1P (partial removal of NAPL in superficial deposits by excavation and off-site disposal) and S2P (partial removal of NAPL in bedrock) are each the most appropriate choice for 18 of the 25 short-listed objectives. Approach S1 (removal of NAPL in superficial deposits and on-site treatment) is most appropriate for 3 of the short-listed objectives.

The analysis can now proceed to step C, a high-level CBA leading to selection of the economically optimal remedial objectives for each of the 11 SPR linkages at the site. It has already been shown that several approaches are likely to figure prominently in the final, complete remedial strategy for the site, but none of the

TABLE 13.19

Cost Analysis of Remedial Approaches

Remedial Approach	Remedial Technology for Costing	Capital Cost $ (£) million	Annual Operation Costs $ (£) million	Operation Time (Yrs)	Assumptions
S1	Complete excavation and treatment of contaminated superficial deposits	1.98 (1.10)	–	–	Soil washing on site + limited landfilling
S1P	Partial excavation of contaminated superficial deposits	0.9 (0.5)	–	–	Material is landfilled
S2P	Partial removal of NAPL in bedrock using angled wells and pumping, water and surfactant flushing	0.9 (0.5)	0.18 (0.10)	10	Sufficient NAPL is removed to have positive impact on dissolved mass flux
P1D	Contain dissolved phase and NAPL in superficial deposits using slurry wall and pumping with treatment	0.54 (0.30)	0.18 (0.10)	20	System must remain operational over long term, shallow source remains in place
P2D	Contain dissolved phase in bedrock by installing hydraulic containment system	0.54 (0.3)	0.18 (0.10)	20	System must remain operational over long term
P4N	Contain vapor at limit of LNAPL plume by installing soil vapor extraction system around site perimeter	0.45 (0.25)	0.09 (0.05)	20	20 yrs without NAPL removal, 2 yrs with
R1	Collect and treat discharge to river at river, using cofferdam at river's edge	1.26 (0.7)	0.18 (0.10)	20	System must remain operational over long term, shallow source remains in place
N3MNA	Monitored natural attenuation of groundwater contamination over time, assuming MNA effective	0.18 (0.1)	0.09 (0.05)	20	Increased monitoring capability

TABLE 13.20
Remedial Technology Category Designations

- *Accepted*: *High probability of success*: Method has been extensively applied under a variety of conditions, is well documented, and has wide acceptance in the industry. Examples: pump-and-treat (for hydraulic containment), slurry walls, air-stripping for VOC removal from pumped water.

- *Emerging*: *Moderate probability of success*: Method has been tested at field scale but is still not fully understood or developed and still requires more detailed field validation. Examples: *in situ* air-sparging (other than for BTEX), funnel-and-gate, natural attenuation, *in situ* bioremediation.

- *Experimental*: *Low probability of success*: Method is undergoing bench-scale or pilot-scale testing, is not well understood, and is currently the subject of significant study. Examples: phytoremediation, *in situ* surfactant washing.

approaches deals with all of the identified risks. Thus, the final overall strategy will involve several approaches.

13.4.6   Step C — Conduct High-Level CBA

At this stage, an economically optimal approach is defined for each objective being considered. The costs and benefits of each objective/approach are compared, and the objective/approach combination that results in the highest net benefit is selected. The costs and benefits are compared through a partial CBA, where the term *partial* indicates that not all identified benefits and costs are monetized.

As discussed earlier, we have found that many of the short-listed remedial approaches actually achieve several objectives. For each approach, benefits are calculated by determining which of the monetized benefit categories are realized. For the purposes of this analysis, aquifer remediation benefits are accrued only when monitoring of the aquifer is undertaken to confirm the results of the remediation. For instance, for approach S1 (complete removal of source in superficial deposits), benefits to the river, property value, and neighboring property are realized. Aquifer benefits are not realized because NAPL remains in bedrock as a source of ongoing dissolved-phase contamination. Approach S2P (partial removal of NAPL from bedrock) accrues no aquifer benefits alone: first, because removal of the bedrock NAPL without removing the source that feeds it in the surficial sediments will result in no net improvement, and second, because it is assumed that benefits of aquifer amelioration cannot be realized or claimed without monitoring (N3MNA) to prove that such improvements are occurring. Thus, for the combination of approaches S1P and S2P, aquifer amelioration benefits are accrued. In contrast, N3MNA alone results in no benefits, because no active remediation is taking place to change the status quo, and the presence of large amounts of NAPL in sediments and in bedrock assures that plume shrinkage is unlikely over time. Table 13.22 shows the benefits apportionment matrix, introduced in Part III, for this example. The proportion of each benefit that would be realized by each remedial approach is provided, with 1 signifying

TABLE 13.21
Probability of Success of Remediation

Remedial Approach	Remedial Technology for Costing	Assumptions	Designation	Probability of Success
S1	Complete excavation and treatment of contaminated superficial deposits	Soil washing on site; fines are land-filled	Accepted	High
S1P	Partial excavation of contaminated superficial deposits — hotspots contributing directly to bedrock contamination	Soil washing on site; fines are land-filled	Accepted	High
S2P	Partial removal of NAPL in bedrock using angled wells and pumping, water and surfactant flushing	Sufficient NAPL is removed to have positive impact on dissolved mass flux in bedrock plume	Emerging to experimental	Moderate to low (technically difficult)
P1D	Contain dissolved-phase and NAPL in superficial deposits using slurry wall and pumping with treatment	System must remain operational over long term, shallow source remains in place	Accepted	High
P2D	Contain dissolved phase in bedrock by installing hydraulic containment system — includes monitoring system	System must remain operational over long term, including monitoring	Accepted	High to moderate (technically difficult)
P4N	Contain vapor at limit of LNAPL plume by installing soil vapor extraction system around site perimeter	20 yrs without NAPL removal, 2 yrs with	Accepted	Moderate (technically difficult)
R1	Collect and treat discharge to river at river, using cofferdam at river's edge	System must remain operational over long term, shallow source remains in place	Accepted	High
N3MNA	Monitored natural attenuation of groundwater contamination over time, assuming MNA is effective	Increased monitoring capability	Emerging	Moderate

TABLE 13.22
Benefits Apportionment Matrix (Present Values)

Approach	Benefit 1 Property value increase ($1.89 million; £ 1.05 million)	Benefit 2 Neighboring properties value increase (blight reduction) ($0.72 million, £0.40 million)	Benefit 3 Improve river water quality ($0.23 million, £0.13 million)	Benefit 4 Improve aquifer water quality (1 = 100% amelioration over 20 years) ($1.49 million; £ 0.83 million)	Sum of benefits ($ million [£ million])
S1	1	1	1	0	2.84 (1.58)
S1+N3MNA	1	1	1	0.1	2.99 (1.66)
S1+P2D	1	1	1	0.5	3.59 (2.0)
S1P	1	1	0	0	2.61 (1.45)
S1P+N3MNA	1	1	0	0.1	2.76 (1.53)
S1P+P2D	1	1	1	0.5	3.59 (2.0)
S2P	0	0	0	0	0 (0)
S2P+S1P+ N3MNA	1	1	0	0.25	2.98 (1.66)
P1D	0	0	1	0	0.23 (0.13)
P2D	0	0	0	0.5	0.75 (0.42)
P4N	0	0.25	0	0	0.18 (0.10)
P4N+S1P	1	1	0	0	2.61 (1.45)
R1	0	0	1	0	0.23 (0.13)
N3MNA	0	0	0	0	0 (0)

TABLE 13.23
Approach Selection — Base Case

Remedial Approach	PV Cost $ (£) Million	PV Benefit $ (£) Million	PV Net Benefit $ (£) Million	Nonquantifiable Benefit Impact	Probability of Success	Risks Managed
S1	1.98 (1.1)	2.84 (1.58)	0.86 (0.48)	++	High	1A, 1B, 3A, 3B, 6A, 7A, 7B
S1P	0.5 (0.9)	2.61 (1.45)	1.71 (0.95)	++	High	1A, 1B, 3A, 3B, 6A, 7A, 7B
S2P	2.29 (1.27)	0	−2.29 (−1.27)	+	Moderate to low (technically difficult)	None without removal of source in sediments
P1D	3.15 (1.75)	0.23 (0.13)	−2.29 (−1.27)	0	High	1A, 1B, 3A, 3B
P2D	2.77 (1.54)	0.75 (0.42)	−2.03 (−1.13)	0	High to moderate (technically difficult)	4 (partial), 5 (partial)
P4N	1.15 (0.64)	0.18 (0.1)	−0.97 (-0.54)	+	Moderate (technically difficult)	6A, 6B
R1	3.51 (1.95)	0.23 (0.13)	−3.28 (-1.82)	0	High	1A, 1B
N3MNA	1.30 (0.72)	0	−1.30 (−0.72)	0	Moderate	5 (partial)

complete realization of that benefit and 0 signifying no realization. To calculate the total benefit realized over the planning horizon for each approach (or combination of approaches), the monetized value of each benefit is multiplied by the factors in Table 13.22 and all benefits are added. So, for approach S1P+P2D, the total PV benefit is:

$$\Sigma \text{ PV (B)} \quad = B1 + B2 + B3 + 0.5 \text{ B4}$$

$$= \$1.89 + \$0.72 + \$0.23 + \$0.75 \text{ million}$$

$$= \$3.59 \text{ million}$$

This is illustrated in Table 13.23, which provides the costs, benefits, and qualifiers for each short-listed approach (the base case, involving immediate start of remediation and a 5% discount rate, is used in this example). The net benefit for each approach is also provided. *Net benefit* is defined as the sum of PV benefits minus the sum of PV costs. If net benefit is negative, the approach is uneconomic. If net benefit is positive, the approach is economic. Among those with positive net benefits, the approach with the highest net benefit is the recommended one on the basis of the economic analysis. The risks managed by each approach are also provided. In some cases, an approach can fully manage a given risk. In others, it will result in partial management of the risk only (it will contribute to management of the risk but cannot do so alone).

The approach of providing safe working practices for on-site workers (R5P) is not considered explicitly, because its low cost would almost guarantee implementation and because it is not strictly related to groundwater questions but is primarily a health and safety measure.

Table 13.23 shows that approaches S1 and S1P manage the greatest number of risks (7 of the 11 identified risks). But approaches P1D, P2D, P4N, R1, and N3MNA satisfy relatively few risks on their own and are also clearly uneconomic if applied alone (all have negative net benefits). N3MNA (institutional management of the aquifer, including the use of monitored natural attenuation, in instances where such effects can be clearly demonstrated) yields no benefit if applied without any form of source removal, because it is not anticipated that any amelioration of aquifer quality will occur. Only the two source-removal approaches (S1 and S1P) are economic when applied alone, but neither is capable of managing *all* of the 11 SPR linkages identified at the site. Clearly, a combination of remedial approaches will be required at this site.

Table 13.24 illustrates the effect of combining approaches in order to manage all or most of the identified SPR linkages at the site. The 14 approaches or combinations of approaches previously identified are considered. It is important to note that benefits arising from aquifer amelioration over time are attributed to remedial approaches that are expected to reduce materially the amount of NAPL in the subsurface, which contributes to the dissolved-phase plume in the sandstone. Because the plume is presently considered to be stable, it is assumed that natural attenuation is presently at work at the fringes of the plume and that the additional mass contributed to the plume by the subsurface sources (in both superficial deposits and bedrock) is removed at the same rate by the active attenuation processes in the subsurface. However, for any economic benefit to occur, an improvement in aquifer quality must occur. Thus, a reduction in the mass flux of contaminant to the plume is required. It is arguable, however, whether these benefits can realistically be included in such analysis without an explicit program of monitoring designed to *prove* that such amelioration is actually occurring. To reflect this, aquifer remediation benefits are *not* included in superficial source removal approach benefit totals, unless monitored natural attenuation (N3MNA) is also included. However, aquifer remediation benefits are included for the two direct bedrock-remediation approaches (S2P and P2D), under the assumption that these activities will include some monitoring of conditions within the sandstone aquifer.

Interestingly, there appears a direct relationship between the overall economic benefit of the MNA approach and the value ascribed to the aquifer. As the value of the aquifer increases, the benefits of MNA rise accordingly and in a linear fashion.

TABLE 13.24
Cost–Benefit Analysis — Approach Combinations — Base Case

Remedial Approach	Description	Complementary Approaches	Benefits PV $ (£) Million	Total PV Cost $ (£) Million	Net Benefit $ (£) Million	BCR
S1	Complete excavation and treatment of contaminated sediments		2.84 (1.58)	1.98 (1.10)	0.86 (0.48)	1.44
		+ N3 MNA — Natural attenuation of remaining plume	2.99 (1.66)	3.28 (2.64)	-0.28 (-0.16)	0.91
		+ P2D — Contain plume in bedrock	3.59 (2.0)	4.75 (1.82)	-1.16 (-0.65)	0.76
S1P	Partial excavation of contaminated sediments (hotspots)		2.61 (1.45)	0.9 (0.5)	1.71 (0.95)	2.90
		+ N3 MNA — Natural attenuation of remaining plume	2.76 (1.53)	2.20 (1.22)	0.56 (0.31)	1.26
		+ P2D — Contain plume in bedrock	3.59 (2.0)	3.67 (2.04)	-0.08 (-0.05)	0.98
S2P	Partial removal of NAPL in bedrock		0	2.29 (1.27)	-2.29 (-1.27)	0.0

+ S1P — Partial removal of NAPL in superficial deposits + N3MNA	2.98 (1.66)	4.48 (2.49)	-1.50 (-0.83)	0.67
PID Contain dissolved phase and NAPL in superficial deposits	0.23 (0.13)	3.15 (1.75)	-2.92 (-1.62)	0.07
P2D Contain dissolved phase in bedrock	0.75 (0.42)	2.77 (1.54)	-2.03 (-1.13)	0.27
P4N Contain vapor at limit of LNAPL plume (site boundary)	0.18 (0.10)	1.15 (0.64)	-0.97 (-0.54)	0.16
+ S1P — Partial removal of NAPL in superficial deposits	2.61 (1.45)	2.05 (1.14)	0.56 (0.31)	1.27
R1 Collect and treat discharge to river at river	0.23 (0.13)	3.51 (1.95)	-3.28 (-1.82)	0.07
N3MNA Monitored natural attenuation	0.0	1.30 (0.72)	-1.30 (-0.72)	0.0

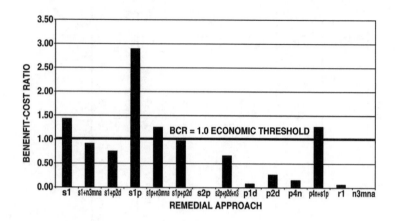

FIGURE 13.1 Benefit–cost ratio for remedial approaches.

In the base case, which assumes that the aquifer will be ameliorated by a maximum of 50% over 20 years and that the replacement value of abstracted groundwater is relatively low (£0.1/m³), total PV benefits from aquifer remediation that can be realized are relatively small, in the order of $0.75 million (£0.42 million) over the 20-year planning horizon. However, if groundwater were more highly valued say $1.8/m³ (£1/m³) and amelioration more successful (say 75% over 20 years), total PV benefits for aquifer amelioration could rise to as much as $11 million (£6 million). This, in turn, could justify increased expenditure on more aggressive aquifer reme- diation techniques and more monitoring of progress.

The results in Table 13.24 are represented graphically in Figure 13.1, which shows the net benefit for each of the 14 approaches and approach combinations examined in the analysis, and in Figure 13.2, which provides the PV costs and benefits for each approach.

Examination of the CBA results in Table 13.24, Figure 13.1, and Figure 13.2 yields some interesting observations. First is that the benefit–cost ratio is maximized when approach S1P (partial excavation of contaminated superficial deposits) is used (BCR = 2.90). This represents the highest ratio of benefits to society (including the problem holder, who in this case achieves a significant benefit from selling a property that has become fit for commercial use) to cost to the problem holder. Using this approach would manage 7 of the 11 SPR linkages outright. However, combining this limited source removal (S1P) with a program of monitored natural attenuation (N3MNA) still provides a positive BCR (1.26). This choice of approach would reflect a decision to consider the distribution of benefits as a key parameter in decision making. In addition, this combination of approaches would address three additional remedial objectives and manage a total of nine risk linkages. Only SPR linkages 2 and 4, both involving the river, would not be satisfied through this solution. However, the associated risk assessment points to the fact that relatively low mass flux to the river, in the form of dissolved-phase contamination, does not result in serious environmental impact. This is due to the poor current quality of the river, situated

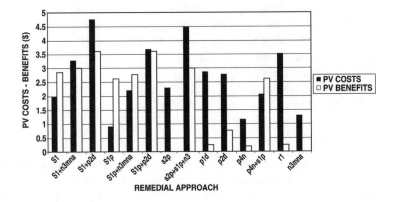

FIGURE 13.2 PV of cost and benefits of remedial approaches.

in a heavily industrialized area and subject to other discharges along its reach. Of course, this situation could change over time, and it is important to note that in the U.K. it is the Environment Agency's objective to improve surface water quality.

Other remedial approaches considered, such as an active attempt to remove some of the NAPL contained in fractured bedrock (S2P), are not economically attractive to society at this site. This approach entails high costs for relatively few overall benefits: BCR = 0 alone, and BCR = 0.67 combined with S1P (hotspot removal and treatment) and monitored natural attenuation of the dissolved-phase plume in bedrock (N3MNA).

The other combination of approaches with a positive BCR is P4N with S1P (containment of organic vapors at the perimeter of the site and partial source removal from superficial deposits). However, this approach combination deals with only the specific concerns of vapor impacts to off-site residents, adding only one additional risk managed (for a total of 8 of 11). In addition, risk assessment has shown that off-site vapor risks are not considered significant.

Using the base case assumptions, economic analysis suggests that excavation and treatment of hotspots of NAPL contamination in the sediments, perhaps combined with some form of monitoring of the dissolved-phase plume in bedrock, is the favored remedial approach. Revisiting Table 13.13, the most economic remedial objective for each SPR linkage is readily determined.

13.5 SENSITIVITY ANALYSIS

Several of the key input parameters in the analysis are subject to uncertainty. It is instructive to examine how variation of these parameters may affect the overall result of the analysis and, ultimately, the choice of remedial objectives and approaches. In particular, variations in the following parameters are examined:

- Discount rate — Typically, a lower discount rate is considered to place a higher value on the conservation of resources in the future and less emphasis on the need to realize positive benefits quickly.
- Timing of remediation
- Value of groundwater
- Mobile plume scenario
- Addition of external costs of remediation

13.5.1 DISCOUNT RATE

The choice of discount rate can significantly affect the outcome of cost–benefit analysis for groundwater remediation or protection. A concern is that discounting effectively devalues the future by putting an inordinate emphasis on present value.

To examine the impact of a lower discount rate on the analysis for the example site, a discount rate of 2% was used. Both costs and benefit flows were discounted, as with the 5% base case.

Figure 13.3 shows a comparison between total benefits in each of the four benefit categories, at 2% and 5% discount rates. Benefits are higher with the lower discount rate for the river and aquifer benefits, where there are flows over time. The total of all benefits over 20 years (fully realized) for an immediate start at 5% is $4.34 million (£2.41 million) (base case) but rises to $5.36 million (£2.98 million) at 2%. However, costs are also discounted. At 2%, the present value of ongoing operation and maintenance (O&M) costs (for approaches where these are required) is also higher than for the 5% base case.

Figure 13.4 shows the BCR of each of the 14 approaches or approach combinations considered, for both the 2% and 5% discount rates. Perhaps surprisingly, only approach S1 experiences an increase in BCR (and thus PV net benefit) with a decline in discount rate. This approach involves capital expenditure in the first year of remediation and no O&M costs. All of the remaining approaches have the same or lower net benefits with the application of a lower discount rate.Examination of

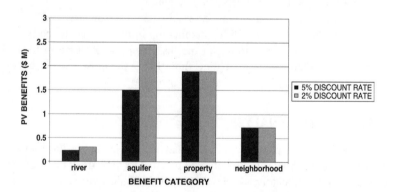

FIGURE 13.3 Sensitivity analysis — present value of benefit with different discount rates.

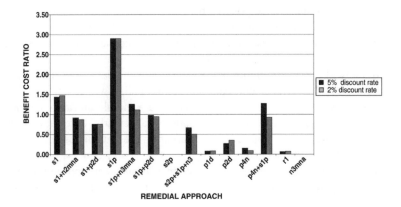

FIGURE 13.4 Sensitivity analysis — benefit–cost ratios with different discount rates.

the benefit and cost flows for these approaches reveals that, in each case, annual O&M costs exceed annual benefit flows. Thus, a lower discount rate actually results in lower PV net benefit. This result is somewhat counterintuitive. Lower discount rates will *not always* result in increased *net* benefits. Higher present value cost profiles will act to lower net benefits — more so in circumstances in which remedial activities require significant ongoing expenditure. Remedial approaches that involve forward-weighted expenditure, however, will tend to realize benefits sooner and incur costs sooner.

Put another way, problem holders with a commercial perspective will be more likely to want to defer expenditure on remediation, all other factors remaining equal, when their PV costs are high and their (private) benefits are low. Many private firms seeking high rates of return on investment capital use high discount rates in their financial analyses. Using a discount rate of 15%, for instance (a not-uncommon target for private sector rate of return), a private problem holder would typically want to defer capital-intensive remediation schemes in favor of ongoing, less intensive approaches. This would particularly be the case if the private benefits of remediation were considered small. In contrast, society as a whole may value the resource being impacted and, applying a lower discount rate to reflect this fact, would place a higher benefit on the resource being damaged, increasing net benefit of remediation and making expenditure on remediation more attractive from the standpoint of the society as a whole. As discussed in Part II and Part III, the choice of discount rate used in the final analysis could be a subject of negotiation among the various stakeholders.

13.5.2 TIMING OF REMEDIATION

Delay or deferral of groundwater remediation is common for many reasons. From an economic perspective, deferring expenditure means that the overall present value of the cost of remediation is lower if the remediation is not seen as providing positive benefits. In many instances, problem holders do not recognize that groundwater remediation may actually have positive benefits, in part because they are not considering the wider social benefits but see the problem only from a narrow, private perspective. In such instances, deferral has obvious financial advantages for the problem holder. This is one of the many reasons for this research and the development of this framework: with a common basis for *all* stakeholders to evaluate the economics of a groundwater problem, better decision making will result.

In cases where contaminants in groundwater are mobile and moving, damage could be increasing over time. In these situations particularly, deferral of remedial action may have significant consequences. Capturing these damages in an economic analysis is vital for good decision making.

Figure 13.5 compares benefits for immediate implementation and a five-year delay in starting remediation, for each remedial approach. Figure 13.6 shows a

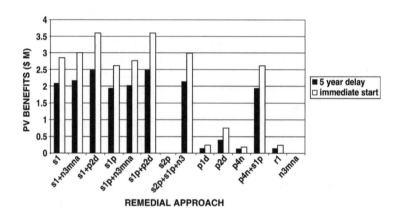

FIGURE 13.5 Sensitivity analysis — present value of benefits with a five-year delay.

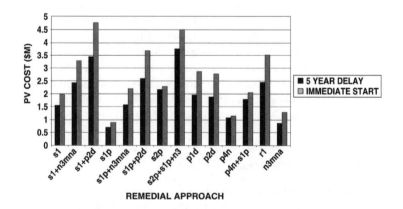

FIGURE 13.6 Sensitivity analysis — present value of costs with a five-year delay.

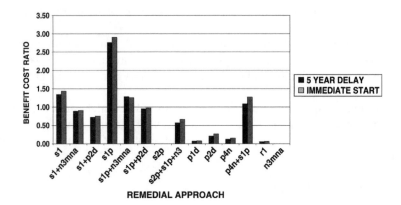

FIGURE 13.7 Sensitivity analysis — benefit–cost ratios with a five-year delay.

similar comparison for the costs of each approach. The clear result of deferral of action is considerably lower PV benefits for all approaches. At this example site, aquifer damages are not increasing with time (the plume is stable). However, deferral of remediation means that benefits from property value increases and improvement in aquifer quality are not realized for some time, and when they are, these are discounted. Costs are also deferred. Present value of the remediation costs also decline when the start date is put back. This effect is more pronounced at higher discount rates.

The effect of a five-year deferral on BCR for each of the 14 remedial approaches is shown in Figure 13.7. BCRs decline due to deferral for all of the 14 approaches except for S1P+MNA. BCR increases with deferral of remediation start-date for S1P+MNA. By examining the cost and benefit flows of each of these approaches, some interesting observations can be made:

- Despite some significant changes in net benefit (as much as one-third), the changes were not significant enough to alter the overall relative ranking of approaches with respect to net benefit.
- Net benefits decline for one of two reasons:
  - For approaches in which expenditure and benefit are one-time, fixed-sum events, closely spaced (S1 and S1P), deferral means that the difference between these values is discounted by five years, when compared to an immediate start. In general, this will apply for approaches with intensive forward-weighted expenditure and more immediate realization of benefits.
  - For approaches where benefit flows over time exceed cost flows or where benefit flows are time-delayed with respect to expenditures, benefits suffer more than costs as a result of deferral.
- For remedial approaches in which the cost flows over time exceed benefit flows over time, and if remedial benefits and costs are accrued preferen-

tially over the longer term, net benefit will tend to *increase* if remediation is deferred.

- In situations in which damage is increasing with time (as with a mobile and expanding plume), delays in remediation are much more likely to mean a reduction in net benefit, all other factors remaining equal.

A further comparison is provided in Figure 13.8, which shows the impact of a ten-year delay on the start of remediation. Delaying remediation by an additional five years does not, in this example, change the relative ranking of the approaches.

13.5.3 VALUE OF GROUNDWATER

One of the major limitations of the benefits analysis conducted for this example site is the lack of reliable and robust studies on aquifer valuation. The value of groundwater has been explored by a number of workers in the past, and some of these studies are discussed in Part II. However, no studies of aquifer valuation are available for the U.K. In the benefits analysis for this site, an approximate valuation of the value of groundwater was developed using the current average sale price of water to domestic users in the U.K., divided by a factor representing the costs of abstraction, treatment, and delivery, and the corporate administration, overheads, and profits of the water companies. A factor of approximately five was used to determine an estimate for the actual costs of the water itself. This resulted in a value of $0.18 (£0.1)/m^3. However, further research is required to determine the average net value that the water companies place on the water itself in arriving at the sale price. This value was then used to estimate the average replacement cost for water rendered unusable by the presence of contamination in the aquifer.

Clearly, alternative valuations for groundwater are possible. Depending on the perspective of the stakeholders, the actual TEV of groundwater (which, by definition,

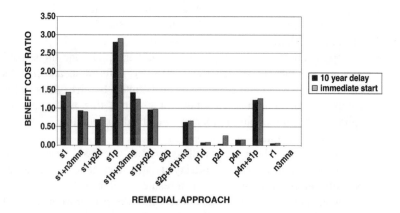

FIGURE 13.8 Sensitivity analysis — benefit–cost ratios with a ten-year delay.

includes use and nonuse values) may vary considerably. Especially if nonuse values are included, estimates of TEV could be substantially higher than the estimate used in the base case of this example analysis.

In an attempt to illustrate the impact of a change in aquifer and groundwater valuation on CBA results, a considerably higher figure of $7.50 (£4.15)/m^3 was used. As before, it was assumed that remediation would result in a gradual improvement in aquifer quality year by year. Figure 13.9 shows the costs and benefits of each of the 14 approaches considered with the alternative aquifer valuation. Figure 13.10 exhibits the BCR of each approach using the base case and alternative groundwater valuations.

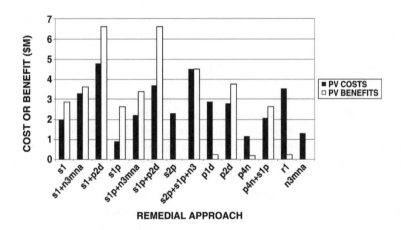

FIGURE 13.9 Sensitivity analysis — present value of costs and benefits with higher value of groundwater.

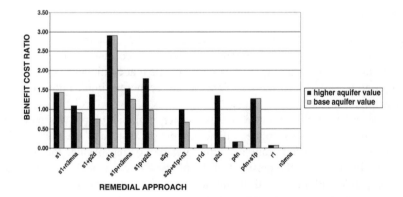

FIGURE 13.10 Sensitivity analysis — benefit–cost ratios with higher value of groundwater.

Approaches that do not generate benefits from aquifer remediation show no change in net benefit. As before, we assume that for a benefit from aquifer improvement to accrue, the problem holder must engage in an approved monitoring program, so the costs of an MNA program are justified in terms of a concomitant benefit. Those approaches that do result in a demonstrable aquifer improvement register significantly higher net benefit as the value of the groundwater resource increases. Substantial increases in net benefit are seen for approaches involving containment of the dissolved-phase plume in bedrock (P2D), monitoring and natural attenuation (N3MNA) of the dissolved-phase plume in bedrock in combination with some attempt at source removal, and remediation of NAPL in the fractured bedrock combined with NAPL removal from overlying sediments (S2P + S1P + N3MNA). All of these move from negative net benefit to positive net benefit as a result of the change in aquifer valuation. These are all approaches that involve a significant effort to restore the aquifer.

The overall effect of aquifer valuation on CBA results is significant. The greater the predicted level of aquifer amelioration and the higher the value of the aquifer are, the larger the benefit of aquifer remediation. Thus, the greater the value placed on the resource is, the greater the remedial effort that can be justified on a CBA basis. Conversely, if a groundwater resource is valued as low or is undervalued, there will be little economic incentive to engage in aquifer restoration or even prevention of future damage. In that case, other benefits, such as the realization of property value increases, will dominate the economic analysis, and the remedial approaches and objectives adopted will reflect this. Similarly, if problem holders are asked to expend significant resources to remediate groundwater with an overestimated TEV, economically inefficient remedial objectives may be set. Without a basic understanding of aquifer valuation, it is unlikely that this issue will be resolved adequately. Application of robust economic analysis for groundwater remediation requires that aquifers be valued accurately.

13.5.4 MOBILE PLUME SCENARIO

The base case assumes that the plume has stabilized at its current position and is no longer expanding. Natural attenuation effects are considered to be active and contributing to plume stability. The case of a plume that continues to expand within the aquifer is considered, to shed light on the economic implications of aquifer damage increasing with time.

As described in the appendix to this chapter, a mobile plume was modeled by considering a conservative solute with no retardation, instead of the benzene plume. The plume was modeled from its birth forward in time. Again, the abstraction potential eliminated by the presence of the plume was calculated, over time, as the plume expanded. Plume growth in this scenario, and thus economic damage (measured as abstraction forgone due to plume presence), was greatest in the first five years and slowed exponentially over time.

To conform with the base case, in which the present is time zero, it can be assumed that the early period of accelerated plume expansion has already occurred at some point in the past. By now, the plume is still growing, albeit at a slower rate. Using the 5- to 30-year plume expansion rate from the modeling, only an additional 25 m³/day of abstraction is lost.

With an expanding plume, the benefits calculated in the base case as a result of aquifer amelioration may not occur. These benefits are lost from the calculation. However, remedial actions that serve to stabilize the plume at its present position result in avoided damages (benefits) equivalent to the annual discounted value of abstraction protected (which, by definition, increases from year to year, to an anticipated steady state at about year 50). For the base case, using a 5% discount rate over 20 years, the benefits of damage avoided by stabilizing a mobile plume would be $0.64 million (£0.36 million). This compares to a benefit of $1.5 million (£0.83 million) for full-plume amelioration over 20 years on a stable plume. In this particular case, the overall effect on the CBA is to increase costs substantially over 20 years (hydraulic containment costs for the plume in its present position are estimated to be in excess of $8.1 million [£4.5 million]) for relatively little additional benefit. Using the alternative higher aquifer valuation, the benefits of aquifer protection increase significantly (as discussed earlier), justifiying the anticipated costs.

13.5.5 EXTERNAL COSTS OF REMEDIATION

In the base case, the external costs of remediation were not included in the analysis. In this example, we assume that all external damages resulting from implementation of remediation are mitigated against, except the cost of off-site disposal of contaminated fines resulting from the soil-washing process. Both S1 and S1P involve this process. In both cases, assuming that 20% of the excavated volume is landfilled and using external costs of road transport of $0.76/vehicle-km (discussed in Part II) over a round-trip distance of 230 km, external costs of remediation of $0.007 million and $0.014 million are calculated for S1 and S1P, respectively. These values are simply added to the cost column. The effect of these costs on the overall analysis is negligible. In this case, on-site soil washing has eliminated much of the road traffic that would have resulted if excavated sediments were sent to a landfill. Remaining volumes are relatively small and the landfill relatively close, so the overall external cost of transport is low. Including these costs does not change the overall economics or conclusions of the analysis.

13.6 UNCERTAINTY AND LIMITATIONS

The example presented herein is subject to inherent limitations and uncertainty, largely the result of required assumptions and the often subjective nature of selections and appraisals that must be made by the user. The groundwater CBA methodology presented in Part III necessarily depends on the expert input of the user in determining which remedial approaches are most appropriate for any given scenario and which risks are most significant and likely. In reality, these are the same limitations inherent in most, if not all, guidance methodologies for contaminated sites: they depend heavily on the expertise and experience of the user and, in some cases, on the perspective of the user or stakeholder. As such, this methodology is seen as a tool for negotiation among stakeholders, each of whom will tend to value various resources and potential risks slightly differently.

Specific limitations of the analysis are:

- Only four main benefit categories are monetized. The effects of other benefits, not readily monetized, are described in a qualitative way. In all of the preceding discussions, it is clear that the benefits of a particular action are at least the monetized value and, in most instances, appreciably above. Refining benefits estimates further than the level presented in this example is unlikely for the majority of sites that are routinely considered in developing countries at present.
- Undertaking the benefit assessment portion of the analysis requires a significant degree of user knowledge and experience. Streamlining and simplifying the benefit analysis to allow greater ease of use should not be at the expense of consistency and robustness.
- Economic value of groundwater in the U.S., the U.K., and most other parts of the world is poorly understood. A greater understanding of this key issue is urgently required if economically optimal decisions involving groundwater remediation are to be made in the future.
- Predictions about the success of aquifer remediation in response to a remedial program are extremely uncertain. In the example, various levels of aquifer amelioration were assigned to various remedial approaches, which impact on CBA results. The greater the value placed on groundwater is, the greater the impact of uncertainty in predicting aquifer amelioration on CBA.
- Property value benefits were modeled as one-time realizations that occurred at the time of sale. In assessing the impacts on the CBA of changing start times, the potential appreciation of property over time was not considered. In certain economic climates, property value may increase appreciably over time, making a remediation deferral advantageous to the property holder.
- The groundwater plume in bedrock was determined to be stable on the basis of modeling results, implying that natural attenuation processes were occurring. This considerably simplifies the analysis. If the plume were actually increasing in size over time, then remediation of groundwater would accrue the additional benefit of preventing additional damage to the aquifer for each year that the damage would have occurred. This could result in the realization of significant benefits. The appropriate valuation of groundwater would become an even more important consideration in this instance.
- Costs of remediation are considered to be stable over time, when assessing the impacts of deferral of start-time. In reality, technology may change over time, as may regulations and law. Indeed, the value of groundwater

as a resource may also change in the future, particularly if the effects of global climate change are considered.

- A 20-year planning horizon was used throughout. Many benefits, such as groundwater improvement, will extend over considerably longer periods of time. Especially at lower discount rates and if groundwater is more highly valued, longer planning horizons may lead to higher present value benefits. However, limitations and uncertainties also exist when using longer planning horizons. Changes in the value of commodities and resources may fluctuate considerably and cannot reasonably be predicted in many instances; regulations, law, and technology may also change, with possibly profound effects on costs and benefits. Under these circumstances, a 20-year planning horizon is considered appropriate for the analysis.

13.7 APPENDIX — GROUNDWATER MODELING

13.7.1 BACKGROUND

An example site was developed to examine issues relating to cost–benefit analysis of groundwater remediation. The example site is based on actual conditions at two sites in the U.K., and the modeling presented here reflects realistic hydrogeological conditions in the Midlands region of the U.K. Site geology and hydrogeology are summarized earlier in this chapter. The site lies next to a river that flows northeast past and immediately adjacent to the site. An area of approximately 20 × 20 m, located in the center of the site, has been contaminated with gasworks-type contaminants, particularly coal tar NAPL, down to a depth of 5 m. Free-phase coal tar has penetrated the Sherwood sandstone below this.

13.7.2 OBJECTIVES

The objectives of the modeling were:

- To develop a contaminant transport model of the hybrid site at "Springfield"
- To test the effects of various different scenarios on the hazard to potential receptors
- To estimate the volume of aquifer rendered unusable (as drinking water) by the contamination

13.7.3 MODEL DESCRIPTION

13.7.3.1 Model Code

MODFLOW, developed by the U.S. Geological Survey, was chosen as the most appropriate software package for this work, because it is capable of representing all the important features of the aquifer system in the area and is generally accepted as the industry standard.[5]

MODFLOW is a modular, three-dimensional, finite difference groundwater flow model, which uses a block-centered, finite difference technique to simulate both horizontal and vertical flows of groundwater in a multilayered system, for steady-state and transient conditions. Abstraction, recharge, and interaction between groundwater and surface water features (e.g., rivers and drains) can also be modeled.

13.7.3.2 Model Parameters

The model measures 7.5×7.5 km and has two layers. These are described in Table 13.25. The base of layer 2 dips from west to east. Layer 1 has a constant thickness of 5 m. The site lies in a valley, next to a river that flows southwest to northeast, at an elevation of approximately 105 mAOD (meters above ordnance datum). Two small hills lie in the southwest and northwest corners of the model, with elevations of over 140 mAOD. Model boundaries are:

- No flow along N, S, E, and W of modeled area in layer 1
- No flow along N, S, and W of modeled area in layer 2
- Constant head of 80 mAOD in E of modeled area in layer 2
- River, which is between 3 and 5 m wide and has an elevation of 115 mAOD in the SW of the modeled area to 80 mAOD in the NE

Note that the middle permian marl is interpreted to be nonleaky and therefore acts as an effective base to the model. The modeled area and boundaries are shown in Figure 13.11.

TABLE 13.25
Model Layers and Hydraulic Parameters

Layer	Description	Thickness (m)	Hydraulic Conductivity (m/day) Kh	Kv	Effective Porosity
1	Made ground and alluvial deposits (in river valley)	5	10	10	0.2
	Sherwood sandstone (beyond river valley)	5	4	0.4	0.2
2	Sherwood sandstone	Varies from 1 m at feather edge in west to 100 m in east and is about 30 m thick at site	4	0.4	0.2

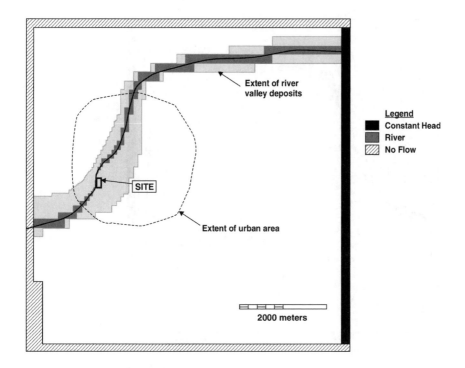

FIGURE 13.11 Model boundaries.

The hydraulic parameters used in the model are given in Table 13.25. Recharge has been set as 250 mm/year in the rural area outside of Springfield and as 51 mm/year in the urban area.

The model has been set up to simulate the transport of benzene, because this is one of the most common risk drivers for a gasworks site. Boundary condition cells with a constant concentration of 30 mg/l have been used in both layers to model free-phase coal tar as a source of benzene in the sandstone and overlying deposits in the center of the site. These cells occupy an area of 20 × 20 m × 5 m deep in layer 1, and 30 × 30 m × 36 m deep in layer 2 (the DNAPL is interpreted to have migrated horizontally, as well as vertically, in the sandstone).

Benzene has been modeled with a K_{oc} of 0.026 m3/kg and a biodegradation half-life of 231 days. There is assumed to be only 0.04% organic carbon in the sandstone and overlying deposits. Dispersion has not been set in the model because numerical dispersion is assumed to represent actual dispersion adequately.

13.7.4 MODEL RESULTS

13.7.4.1 Model Representation of Hybrid Site

Modeled piezometry for run 1 is shown in Figure 13.12. This shows that regional flow is predominantly toward the constant head boundary in the east of the model,

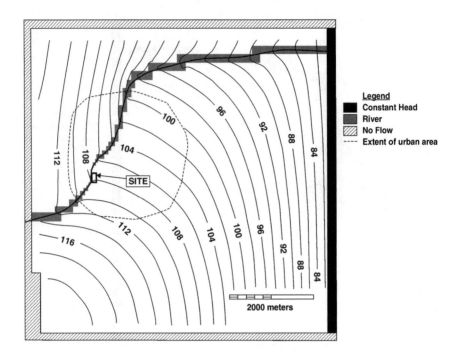

FIGURE 13.12 Modeled groundwater surface elevations (mOAD) — run 1.

although a proportion of flow is toward the river. The water balance for the model shows that 37% of the recharge enters the river and that 63% discharges to the constant head boundary.

Forward particle tracking from the source area shows that groundwater flows both north and northwest to the river in layer 1, and north to the river in layer 2 (Figure 13.13). Modeled steady-state benzene concentrations in layer 1 and layer 2 are shown in Figure 13.14 and Figure 13.15, respectively.

The mass flux of benzene to the river reaches a steady-state value of 3.5 g/day after 3,000 days (Figure 13.16). The hazard to the river will depend on the river's flow rate, because this will affect the benzene concentration in the river water. For a river flow rate of 10,000 m³/day (within the typical range for a river 3 m wide) the benzene concentration within the river will be 0.35 µg/l. The freshwater U.K. standard for benzene is 30 µg/l.

13.7.4.2 Additional Off-Site Source (Run 2)

A leaking UST containing petrol was modeled 120 m south (up hydraulic gradient) from the site. The petrol was modeled as giving a 50 × 40 m area with 30 mg/l benzene in layer 1 and a 50 × 40 m area with 5 mg/l benzene in layer 2. Results show that benzene from the leaking UST will reach the site, but that it is mostly biodegraded by the time it reaches the river and therefore makes little difference to the mass flux of benzene entering the river (3.55 g/day as opposed to 3.5 g/day for cba1).

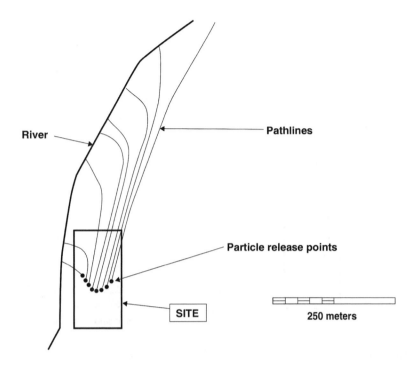

FIGURE 13.13 Pathlines from source area at site (run 1).

13.7.4.3 Non–Hard Covered Site (Run 3)

Recharge has been increased to the rural value of 250 mm/yr at the site. This makes little difference to the groundwater flux occurring beneath the site and therefore has little impact on the benzene flux to the river (3.7 g/day as opposed to 3.5 g/day for cba1).

13.7.4.4 Reduced Effective Porosity (Run 4)

The effective porosity of the sandstone in layer 2 has been decreased from 0.2 to 0.02. This decreases transit times and therefore decreases the amount of benzene lost through biodegradation. Total flux of benzene to the river is now 28 g/day.

13.7.4.5 Reduced Vertical Hydraulic Conductivity of Sandstone (Run 9)

The vertical hydraulic conductivity of the sandstone was increased to 4 m/d. This run is not one of the scenarios in the hybrid site report but was conducted to examine the sensitivity of this parameter on the model results. Increasing the vertical hydraulic conductivity of the sandstone allows more of the total model groundwater flux to occur in layer 2. The constant head boundary in the east therefore has a greater influence on groundwater flow direction in both layers at the site. Groundwater at the site now flows north rather than northwest (Figure 13.17). Although it still

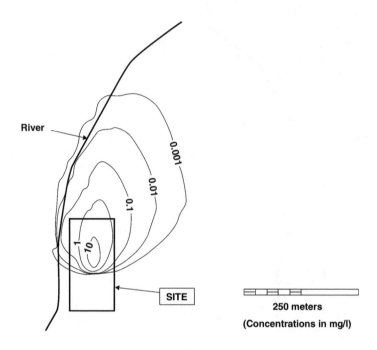

FIGURE 13.14 Modeled groundwater benzene concentrations in layer 1 (run 1).

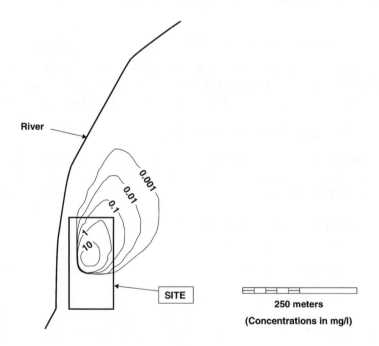

FIGURE 13.15 Modeled groundwater benzene concentrations in layer 2 (run 1).

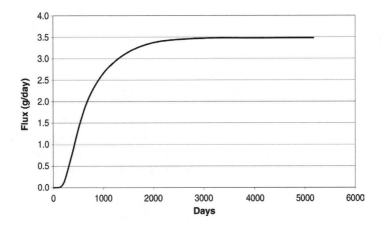

FIGURE 13.16 Modeled flux of benzene to river (run 1).

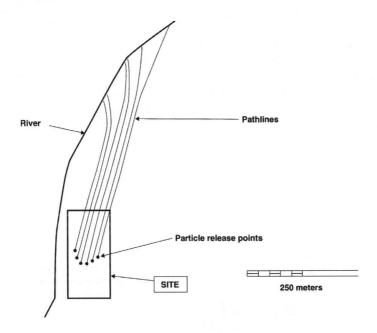

FIGURE 13.17 Pathlines from source area at site (run 9).

discharges to the river, it takes longer to get there, allowing more biodegradation to occur. Benzene flux to the river is now only 0.035 g/day (i.e., two orders of magnitude less than run 1).

13.7.4.6 PWS Well Close to Site (Run 10)

A public water supply (PWS) well abstracting at 5000 m³/day was modeled 500 m to the east of the site. Groundwater from the site now flows toward the PWS well and not into the river. A steady-state flux of 532 g/day benzene to the well is reached after 1500 days. Concentration within the well would therefore be 100 mg/l, which is well over the EC drinking-water guideline of 1 µg/l.

13.7.4.7 Volume of Aquifer Rendered Unfit for Drinking-Water Abstraction (Runs 11, 16, and 17)

Capture zone modeling was used to predict the locations of a PWS well that would result in its abstracting water from the contaminated part of the aquifer. The contaminated part of the aquifer was defined as the steady-state benzene plume derived from model run cba1. An abstraction rate of 5000 m³/d (typical of PWS wells) was used for the abstraction well.

Plotting the locations of the PWS well that resulted in capture of water from the contaminated plume allowed an estimate to be gained of the unusable area of the aquifer. Note that this area is dependent on the abstraction rate of the PWS well. Use of a lower abstraction rate would result in a lower area of unusable aquifer. The predicted area of unusable aquifer using a PWS abstraction rate of 5000 m³/d is shown in Figure 13.18.

The total unusable area is predicted to be approximately 1200 ha, 46% of which is rural land. The total recharge to this area is estimated to be 4607 m³/d. This is the estimated loss in total yield caused by the steady-state benzene plume and assuming an abstraction rate of 5000 m³/d.

Further model runs were conducted to predict the increase in unusable area with time due to plume growth. These runs were conducted for an ammonium

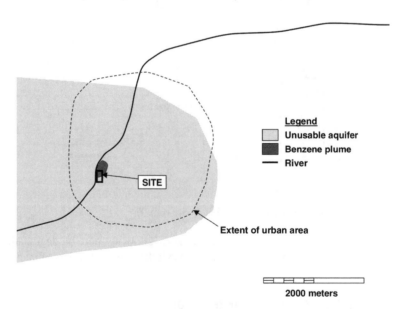

Legend
Unusable aquifer
Benzene plume
—— River

SITE

Extent of urban area

2000 meters

FIGURE 13.18 Predicted unusable area of aquifer from benzene contamination with 5000 m³/d abstraction (run 11).

plume because the size of the benzene plume was insufficient to show a significant change in unusable area with time. Ammonium is another common gasworks contaminant that is not easily degraded. Model run 16 was used to predict the size of the ammonium plume after 5, 10, and 20 years and after steady-state conditions had been reached. The same source area was used as run 1, with a source concentration of 100 mg/l. Ammonium was modeled as nonretarded and nondegradable (a conservative solute). Steady-state conditions are reached after approximately 50 years.

The total yield lost due to the ammonium plume is shown with time in Figure 13.19. This has been calculated using the same method as run 11 but with a PWS abstraction rate of 2500 m³/d. This shows that a rapid loss in yield occurs in the first five years of the source's presence.

13.7.4.8 Effects of Remediation (Runs 12 to 15)

The effects of remediation were modeled for various remedial options, using model run 1 as the initial conditions.

13.7.4.9 Shallow Source Removal (Run 12)

Removal of the source from the surficial deposits leads to a slight reduction in plume size in these deposits. This remediation does not significantly alter the size of the plume in the sandstone.

13.7.4.10 Shallow Source Removal and Source Reduction in Sandstone (Runs 13 and 14)

Removal of the source from the surficial deposits and a reduction in source concentration to 15 mg/l results in a decrease in plume area in the sandstone from 33,000 to 28,700 m². Reducing the source concentration in the sandstone to 5 mg/l results in the plume area's decreasing to 22,400 m².

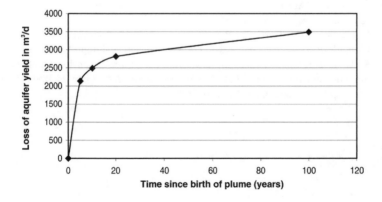

FIGURE 13.19 Predicted loss of aquifer yield with time (run 17).

13.7.4.11 Complete Removal of Source (Run 15)

Complete removal of the source term results in no detectable concentration (< 0.001 mg/l) of benzene off site in either the sandstone or surficial deposits after 11.5 years.

13.7.5 DISCUSSION

The modeling work conducted gives a good indication of the likely outcome of the different scenarios tested with respect to the hazard from benzene contamination. It should be noted that the effect of each scenario on hazard from other contaminants could be different from those described previously, depending on their properties. For example, the modeling conducted shows that hazard due to benzene is dependent on the contaminant travel time between source and receptor. The hazard from contaminants that are not as easily biodegraded, such as cyanide or ammonium, will be less dependent on the travel time.

Several of the scenarios discussed in the main example have not been tested: most notably, vapor hazard to off-site receptors from groundwater contamination and the introduction of groundwater compliance points. The existing model results can be used to examine hazard from these scenarios. Vapor concentrations at a receptor can be calculated using simple one-dimensional vapor models (such as Farmer's model) with the groundwater concentrations predicted by the groundwater modeling. The effect of compliance points can also be easily examined using the contaminant concentration contours given in the preceding figures.

REFERENCES

1. Komex and eftec, *Costs and Benefits Associated with Remediation of Contaminated Groundwater: A Framework for Assessment*, U.K. Environment Agency Technical Report P279, Bristol, 2000.
2. Mendoza, C.A. and McAlary, T.A., Modeling of groundwater contamination caused by organic solvent vapors, *Ground Water*, 28, 2, 199, 1990.
3. Environment Agency, *Policy and Practice for the Protection of Groundwater*, Environment Agency, Bristol, 1998.
4. Komex and eftec, *Costs and Benefits Associated with Remediation of Contaminated Groundwater: A Review of the Issues*, EA R&D Technical Report P278, Bristol, 1999.
5. McDonald, M.G. and Harbaugh, A.W., A modular three-dimensional finite difference ground-water flow model, *U.S. Geological Survey Techniques of Water Resources Investigations*, book 6, ch. A1, 586, 1988.

14 MtBE-Contaminated Aquifer in the United States

The case history presented in Chapter 13 involved a fairly comprehensive and detailed analysis of the costs and benefits of a number of remedial approaches, with multiple source–pathway–receptor (SPR) risk linkages and several possible remedial approaches being evaluated. Chapter 13 also followed most, if not all, of the steps in the evaluation framework presented in Part III.

The following case history illustrates a more preliminary and simplified application of a CBA to a groundwater problem. In many real situations, a less detailed look at costs and benefits will be justified. In these cases, some parts of the framework may be used and others dispensed with, depending on the data available, the funds allotted for decision-making and economic analysis, and the scale of the problem at hand. Even basic evaluations of costs and benefits can provide valuable insight in how best to manage a groundwater contamination situation.

14.1 SETTING

A release of MtBE-enriched gasoline has occurred from an underground tank, contaminating the subsurface. Gasoline entered the thin, permeable surficial sands and migrated into the underlying fractured sandstone bedrock. The groundwater surface at the time of the spill lay slightly below the bedrock–surficial sediments interface, and LNAPL penetrated several meters into the bedrock via large open vertical fractures. The extensive bedrock aquifer is used for domestic and municipal water supply throughout this rural community. Despite rapid excavation and treatment of easily accessible contaminated sand at the release point, subsequent monitoring of nearby wells soon revealed the presence of low µg/L levels of MtBE. Hydrogeological characterization revealed that groundwater flow in bedrock is to the north toward a major river that lies about 5 km away. If the bedrock is treated as an equivalent porous medium (EPM), average groundwater flow velocities of about 250 m/year are estimated. Between the release site and the river are approximately 50 residences with their own water wells.

14.2 BENEFITS OF REMEDIATION — AVOIDANCE OF EXPECTED DAMAGES

A simple analysis of the MtBE plume's migration over time was completed, tracking the plume as it moved toward the river. Because MtBE is a relatively conservative

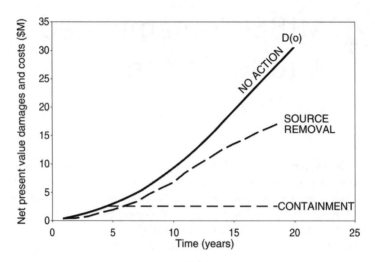

FIGURE 14.1 Remedial benefit (as damage avoided).

solute and does not degrade over time, dilution and dispersion are the main active attenuation mechanisms. Although MtBE concentrations are expected to decrease over time, experience has shown that the public reacts adversely to even low ppb levels in drinking water (due to distinct odor and taste). Damages are estimated based on two components:

- Damage to wells impacted by MtBE, estimated as the cost of supplying water to residents ($25,000 per year per well)
- Notional estimate of the overall economic impact of aquifer degradation, assessed as development of an alternative water source ($10,000 per year per hectare of affected aquifer)

A discount rate of 5% is used. This produces an expected damage curve (D(o)), shown in Figure 14.1. Note that without remedial action, and even with discounting, the damages rise quickly, as the plume expands and more receptors are impacted.

14.3 REMEDIAL APPROACH ANALYSIS

For simplicity, we assume that the objective of remediation is to prevent unacceptable impacts on residents who have not yet been affected by the plume. Because MtBE is not predicted to be present in levels harmful to human health but will be present to the degree that taste and odor are affected, economic analysis of remedial approaches focus on these latter impacts. We consider three remedial approaches:

- Remove the source of MtBE by removing gasoline LNAPL from surficial sediments and bedrock, allowing the existing dissolved phase plume to attenuate over time.
- Contain the plume at its present position.
- Remove the source and contain the existing plume.

For ease of analysis, we consider that all remedial actions are taken in year 2.

Figure 14.1 shows the expected damage curves for the three remedial approaches. If the source is removed, the existing plume moves away over time, affecting wells for a time before the "slug" moves past and the well returns to a more-or-less pristine state. This reflects one of the positives of MtBE contamination — it does not tend to adsorb to aquifer material, so aquifers can be largely rid of MtBE simply through natural flow or extended pumping. However, it takes over 20 years for the plume to work its way through the aquifer and eventually discharge into the river. If containment were put in place immediately, the damage curve would be flat after year 2. Clearly, the benefits from this action would be considerable ($27.5 million over 20 years). The benefit of source removal is $12.5 million over 20 years reflecting a reduction (but not an elimination) of MEBE flux to receptors. The source removal plus containment option shows no difference in benefits within the planning horizon — both achieve the goal fully. However, as discussed subsequently, there are cost differences between these approaches that manifest themselves in both the short and the long terms.

14.4 COST–BENEFIT ANALYSIS

Table 14.1 shows the costs for the containment and source removal plus containment options. Source removal alone is assumed to involve a one-time lump-sum cost of $7.5 million (reflecting the difficulties in bedrock remediation). Costs for containment are based on an initial capital cost of $1.5 million, with an annual O&M cost of $0.25 million and phased capital upgrades of $0.5 million at years 6 and 13. All costs are discounted at 5%.

TABLE 14.1
Remedial Costs

Year	Discounted Cost ($M) Containment (Capital + O&M)	Discounted Cost ($M) Source Removal + Containment (Capital + O&M)
1	1.75	9.25
5	2.59	10.09
10	3.81	11.31
20	5.26	12.76

When the costs are overlain with the damages (benefits curves) in Figure 14.1, some interesting points are revealed. Over a 20-year planning horizon, the net benefit of the containment option alone is about $22.2 million ($27.5 million in benefits minus $5.26 million in costs), but clearly the system must continue to operate over an extended period into the future to prevent damage. Over the same period, source removal has a net benefit of about zero (break-even: $12.5 million in benefits minus $12.76 million in costs), although residual damages that may occur as a result of this strategy have not been taken into account (e.g., real or perceived impacts on human health and river ecosystems). Combining source removal and containment results in a net benefit of about $9.7 million over 20 years. The major advantage of this combined approach, which is not reflected within the 20-year planning horizon, is that the containment system could eventually be shut down. With the NAPL source still present with the bedrock, it is likely that a containment system would have to continue running for much longer than 20 years. In this case, all other conditions remaining the same, the combined approach would become more economic and the containment option less so. However, the effects of discounting play heavily in this analysis. Note that the present value of spending $7.5 million in 20 years' time is only $2.83 million. Thus, it may be advantageous for many firms to consider deferring the more expensive source removal to some point in the future, accruing the immediate benefits of receptor protection through containment only. This is an example of a phased remediation based on economic considerations.

14.5 SUMMARY — DECISION MAKING

In our example, a clear remedial objective is selected at the outset — protection of water wells nearby. Net benefits are maximized (in the 20-year planning horizon) when an approach of containment is adopted. The high costs of source removal in difficult bedrock conditions significantly impact net benefit within this planning horizon. However, it is important to note that beyond this planning horizon, the benefits of source removal become apparent. The option that relied heavily on natural attenuation of dissolved-phase plumes (source removal only) yielded much lower net benefits, primarily because of the recalcitrant nature of MtBE. The effects of discounting are also seen and are a point worthy of serious debate. That this is a partial cost–benefit analysis is also apparent. Other, less-readily monetized benefits will exist and will tend to push up the value of remedial action.

15 Tritium-Contaminated Groundwater

15.1 BACKGROUND

Investigation work at an operational nuclear facility site has shown that concentrations of radioactive tritium in groundwater in some areas of the site exceed background levels. The site owner wishes to consider various remedial options for dealing with tritium-contaminated groundwater at the site, including the objective of remediating shallow groundwater to background tritium levels (50 Bq/l). Another possible objective would allow tritium discharge at concentrations above the background level but below current World Health Organization (WHO) drinking-water standard concentrations for tritium.

15.1.1 SITE DESCRIPTION

The site landfill (approximately 3.5 hectares in area) is located in the central portion of the site. A small stream lies about 500 m to the south of the landfill. The majority of the site is hard-surfaced, the main exception being the site landfill. Surface drainage from the entire site is routed to a collection system and then discharged to the stream.

15.1.1.1 Geology

The geology generally consists of unconsolidated sands and gravels overlying a thick sequence of clay. Across the site, the sand and gravel layer varies in thickness from less than a meter to over 11 m. The upper part of the clay layer consists of a firm-to-stiff laminated clay, typically 1 to 4 m thick. This is underlain by very clayey fine sand with thickness ranging from 0 to 5.2 m. Below that, the clay is stiff brown and dark blue-gray silty clay, with a minimum thickness of 30 m. Underlying the clay at a depth of more than 100 m is a major carbonate aquifer.

15.1.1.2 Hydrogeology

Although perched water is likely to be present within the site landfill, data suggest that it is isolated in nature, rather than a continuous groundwater-bearing unit. Groundwater flow in the sand and gravel unit is toward the northwest, in the direction of the stream. Piezometric heads vary across the landfill from 100 to 90 m above datum. The closest groundwater supply well is completed within the carbonate aquifer, more than 10 km away from the site. There are no supply wells within 2 km completed within the sand and gravel unit. There are no direct water withdrawals from the stream.

15.1.2 Assessment of Contamination

This assessment focuses mainly on tritium concentrations across the site. In the area of the site landfill, other contaminants are also described.

15.1.2.1 Source of Tritium

Tritium has been used in numerous process areas throughout the site's history. Tritium is currently used and stored at site. No leaks have been identified from current operations. Landfilling activities at the site commenced several decades ago, and tritium was among the compounds disposed of in the landfill.

15.1.2.2 Groundwater

Monitoring results suggest that groundwater entering the site from the west has concentrations of tritium averaging 10 Bq/l. Monitoring results from across the whole site (with the exception of the landfill and vicinity) indicate that tritium concentrations generally range from 10 to 50 Bq/l. Tritium concentrations significantly exceeding 50 Bq/l have been found down-gradient of the site landfill. Investigation down-gradient of the landfill has identified a narrow plume with tritium concentrations up to 2500 Bq/l. Groundwater samples from monitoring wells within the landfill showed concentrations of up to 2000 Bq/l.

Elevated tritium concentrations in the site landfill area are generally restricted to the sand and gravel unit. Monitoring wells completed in the clay unit indicate that tritium levels are generally in the range of 10 to 50 Bq/l.

15.1.2.3 Soils

For boreholes drilled within the landfill, tritium concentrations in soil were generally less than 0.4 Bq/g, with a maximum concentration of 1.1 Bq/g.

15.2 RISK ASSESSMENT

15.2.1 Health Effects from Tritium Exposure

Tritium is a radioactive form of hydrogen. Tritiated groundwater (HTO) results from the exchanging of one of water's hydrogen atoms for a tritium atom. Tritium may be produced from anthropogenic sources (i.e., nuclear reactions and biomedical research) or naturally through interaction of cosmic rays with the atmosphere.

The half-life of tritium is 12.3 years. Radioactive decay transforms it into nonradioactive helium through emission of a beta particle. The energy radiation from tritium is too low to be a health hazard to humans if the tritium is present outside the body. A radiation dose can be delivered if tritium is taken inside the body, tritiated water being the most likely source.

There are no human toxicological studies that show health effects due to exposure to low-level radiation. All guidelines are based on the assumption that there *is* a low-dose effect. However, these health risk estimates are extrapolated from high-dose studies.

Drinking-water limits of 740 Bq/l and 7800 Bq/l have been set by the USEPA (U.S. Environmental Protection Agency) and WHO, respectively. The Canadian drinking water guidelines for tritium are currently 40,000 Bq/l, but there is a proposal to revise it to 7000 Bq/l based on the WHO standard. On this basis, this risk analysis assumes that no health risk to humans currently exists, in relation to groundwater discharge to streams, rivers, or sewage works in the area.

15.2.2 SOURCE ANALYSIS

The site landfill appears to be the main source of tritiated groundwater on the site. Several smaller tritium groundwater sources exist up-gradient of the tip, but concentrations down-gradient of these sources are about 50 Bq/l (background) by the time they reach the landfill area. It is thought that the source is present within the landfill as several small solid tritium sources of no more than about 1 g. Rain washout from stack discharges is not likely to contribute significantly to groundwater levels at the site.

15.2.3 PATHWAYS AND RECEPTORS

Tritium concentrations in groundwater the clay are not significant. The highest recorded levels are consistently found in the sand and gravel unit within and down-gradient of the landfill. Based on available data and using the results from a calibrated groundwater flow model generated for the site landfill, the hydraulic conductivity for the sand and gravel unit is estimated at 8 m/day.

The variability of tritium concentrations between sampling rounds suggests that the tritium source may be located near the groundwater surface, with high concentrations occurring when the groundwater levels rise and the source becomes saturated.

The following receptors have been identified at the site:

- Stream — The stream receives tritium via discharge from shallow groundwater. The concentration of tritium received by the river is 30 to 40 Bq/l; however, this is rapidly diluted within the stream.
- Gravel aquifer — Although the nearest abstraction is 2 km north and is unlikely to be affected, the aquifer at the site can be viewed as a future potential source of groundwater that requires protection.

The risk assessment identified two likely source–pathway–receptor (SPR) risk linkages that exist in connection with tritium-contaminated groundwater at the site. These are listed in Table 15.1.

15.3 REMEDIAL OBJECTIVES AND APPROACHES

15.3.1 METHOD

As discussed in Part III, the possible remedial objectives for any SPR linkage are limited to the categories in Table 15.2.

TABLE 15.1
Risk Linkages

	Receptor	Pathway	Source
1	Stream	Discharge from groundwater	Tip area
2	Gravel aquifer and users	Via plateau gravels	Tip area

TABLE 15.2
Remedial Objective Categories

A	Protect potential receptor from future impact
B	Reduce impact to current receptor
C	Eliminate impact to current receptor
D	Remove all contamination
E	Remove contamination to set level
F	No action/institutional management

In this case, the aquifer and the stream are defined as current receptors. Reducing impact to the stream (reducing influent mass flux of tritium) would be objective B. Eliminating all tritium flux to the stream (essentially meeting the objective of background quality) would be objective C. Objective F would be maintaining the status quo of discharge, under a managed regime of careful monitoring. Actions to remove the sources of tritium in groundwater would classify as objectives D and E, depending on the degree of removal. All objectives are worthy of analysis in this case.

15.3.2 CONSTRAINTS ANALYSIS

It is assumed that there are no time constraints on remedial activity at the site. Any anticipated groundwater remedial activity can take place over an extended period without hampering site activities. There are no significant physical constraints in terms of source and pathway because the source and immediate down-gradient area are on site and accessible. In relation to receptor constraints, the stream is located outside of the property boundary, making direct remedial activity within the stream potentially difficult from legal, access, and permitting points of view.

15.3.3 SHORT-LIST OF REMEDIAL APPROACHES

Each remedial objective can be achieved in a number of different ways, using a number of different technologies, applied in various combinations at different places and times. Together, these variables are used to describe the remedial approach. For example, the objective may be to protect the stream from tritium discharge via shallow groundwater. This could be achieved by removing the source, by collecting all groundwater just as it discharges into the stream, by changing the hydraulic gradient in the aquifer (hydraulic containment), or by diverting the stream entirely.

Remedial approaches focus on how the SPR linkage is to be broken to achieve the specified objective. Benefits could vary, depending on the remedial approach selected. Thus, the benefits part of the economic analysis requires that an objective be coupled with an economically optimal approach. Approach selection will also have a direct impact on cost, because different approaches will typically involve different technologies and thus varying costs. Determining which approach is best means that alternative approaches must be considered for each objective and the costs and benefits of each evaluated at a high level. Only then can the most economic objective for a given SPR linkage be determined. Approaches can be generally categorized as in Table 15.3. A list of available remedial approaches is presented in Table 15.4.

Although "no action" is unlikely to be an acceptable approach in itself, this objective is analyzed because it forms the base case and provides a means of assessing the range of costs and benefits over a broad range of approaches. This is particularly useful in this case because the tritium concentrations in groundwater do not pose a public health risk, but their release to the environment may be considered undesirable by some stakeholders due to tritium's radioactive nature.

Relative overall cost, advantages, and disadvantages for each approach are also considered in Table 15.4. At this point in the analysis, consideration of relative costs provides a precursor to development of high-level cost functions for each approach. By using relative cost, the short-list of applicable approaches for further analysis can be further reduced, helping to simplify the analysis. Note that in all, only six remedial approaches have been found worthy of more detailed analysis by CBA for this site. Using a receptor approach for the stream has not been evaluated due to the physical constraints of working off site.

TABLE 15.3
General Remedial Approaches

Remedial Approach	Option	Example
Source removal	Complete removal	Excavation of all source contamination
	Partial removal	Excavation of contamination to set level
Pathway elimination	Complete elimination	Cut-off wall or hydraulic containment
	Partial elimination	Pumping groundwater to remove contamination to a set level
Receptor isolation	Full protection of receptor	Protective working practices
	Partial protection	Protection to set level
No action	Institutional management	Monitor natural attenuation processes

TABLE 15.4
Remedial Approach Alternatives

Designation	Approach Description	Relative Cost	Disadvantage
	Source Methods		
S1	Remove solid tritium from site landfill (remove full volume of landfill)	Very high	Disturbance due to large-scale works, difficulty in disposal
S1P	Partial removal of solid tritium from site landfill (minimize removal through further SI)	Moderate to high	Postremediation monitoring required
	Pathway Methods		
P1	Capture and treat tritiated groundwater from gravel aquifer down-gradient of site landfill	Moderate to high	Long-term operation and maintenance (O&M) required
P2PR	Phytoremediation of tritiated groundwater *in situ*	Moderate to low	Considerable R&D required; considerable risk of failure; long-term O&M needed
	Receptor Methods		
R1	Capture and treat tritiated groundwater before it enters the stream	Moderate to high	Aquifer remains contaminated
	No Action		
MNA	No action; monitor concentrations	Low to moderate	Status quo may not be acceptable to regulators

The possibility of removing the source of tritium from the landfill (approach S1 to meet objective D) has been considered for the purpose of completeness. Considering the economics of such objectives/approaches is useful insofar as it puts other alternatives into context. In reality, however, the source of tritiated groundwater could be in the form of only a few grams of solid tritium. Isolating such a small volume within the entire landfill mass would be quite difficult, so a large proportion of the landfill waste would likely require removal. A statistically driven grid sampling program using CPT (cone penetrometer) and other push technologies could narrow the uncertainty considerably, allowing a more limited removal. Validation could be accomplished through screening of excavated material and down-gradient groundwater monitoring.

Specific remedial objectives and corresponding approach options for SPR link-ages are summarized in Table 15.5. If one particular remedial approach is the clear choice for achieving the objective, due either to the particular physical or environmental considerations at the site or to specific legal or regulatory requirements for action, it is selected. In this case, the financial and external costs of the various approach options are similar, and it is not possible to discount any of them until higher-level cost functions are applied.

On this site, the analysis is considerably simplified by the fact that there is only one significant source–pathway scenario, which applies to all potential receptors. Benefits vary depending on which remedial objective is achieved, not how it is achieved. The role of the CBA is now to assess the costs and benefits of each approach–objective combination. This helps to answer the first of two important appraisal questions: "Is the objective worth achieving?" The second question — "If so, how?" — refers, in this case, to remedial approach options.

15.4 HIGH-LEVEL BENEFITS ANALYSIS

15.4.1 CONTEXT

Some stakeholders are exerting considerable pressure to show some level of action on this issue at the site. However, the low recorded levels of HTO (tritium) in groundwater and the lack of any definable potential for health effects make it difficult to justify potentially expensive remedial actions.

On this basis, the benefits of remediation must be considered carefully. The approach taken for this analysis is to attempt to capture the maximum likely remedial benefits that would accrue to the problem holder and to society as a whole, should

TABLE 15.5
Remedial Approach Options

SPR Linkage	Remedial Objective Options	Remedial Approach Options for Each Objective
1A. Discharge to stream via groundwater	B	S1, S1P, P1, P2PR, MNA, R1,
	C	S1P+P2PR
	D	R1, P1, S1P+P1
	E	S1
	F	S1P, S1P+P1, S1P+P2PR
		MNA
2. Contamination of the gravel aquifer as a current or future receptor (including future users)	B	S1, S1P, P1, P2PR, S1P+P2PR, S1P+P1
	C	S1P+P1
	D	S1
	E	S1P+P1, S1P+P2PR
	F	MNA

TABLE 15.6
Receptor Risk Information

	Receptor	Pathway	Contaminant Impacting	Expected Impact
1	Stream	Discharge from groundwater	Tritiated groundwater	Tritium flux diluted in river. Tritium concentrations not at a level that would harm people or the environment. Loss of amenity due to perceived hazard. Potential blight on property value in areas near stream, downstream of site.
2	Gravel aquifer and users	Via plateau gravels	Tritiated groundwater	No nearby abstractions at present. Loss of option value of aquifer.

various remedial objectives be reached. To do this, the approach adopted uses the largest conceivable value for each monetizable benefit. In addition, a qualitative examination of some likely nonmonetizable benefits is also included. Thus, in the cost–benefit analysis, likely costs are compared with upper-bound benefits. In adopting this approach, the economic analysis is biased toward the position of the regulator and other external stakeholders. What results, in effect, is a worst-case economic analysis for the problem holder.

15.4.2 LINK WITH RISK ASSESSMENT

The results of the risk assessment were used previously to assist in selection and short-listing of remedial objectives. At this stage, the risk assessment is used to provide specific information about identified receptors, the degree of current and anticipated impact, and the relative importance placed on those receptors (Table 15.6). This, in turn, is used to assist in valuing the damage to those receptors, which is translated into a monetized benefit from damage avoided or repaired.

15.4.3 POSSIBLE BENEFITS OF EACH APPROACH

Using the information from the risk assessment, possible benefits are identified that may accrue from successful elimination of an SPR linkage. Possible financial (private) benefits to the problem holder are listed in Table 15.7. Possible wider economic benefits (external) are listed in Table 15.8. Benefits are classified as very likely (VL), likely (L), or not likely (NL), based on how probable the realization of that benefit is if the SPR link is broken.

In both tables, we need only consider those benefits that are likely (L) or very likely (VL) to be realized. What is apparent from these two tables is that the analysis quickly reduces to a few key benefits that are likely to be realized:

TABLE 15.7
Possible Financial (Private) Benefits for All Approaches

SPR	Receptor	Possible Benefits- Financial Analysis	Likelihood of Realizing Benefit
1	Stream	Reduce corporate liability	L
		Reduce possibility of litigation or fines	L
		Positive public relations value	NL
2	Gravel aquifer and users	Reduce corporate liability	L
		Reduce possibility of litigation or fines	L
		Positive public relations value	NL

TABLE15.8
Possible Economic (External) Benefits for All Approaches

SPR	Receptor	Possible Benefits — Economic Analysis	Likelihood of Realizing Benefit
1	Stream	Avoid perceived damage to stream	L
		Perceived improvement in recreational value of stream	L
		Increased property value in neighborhood due to blight elimination	L
2	Gravel aquifer and users	Restoration of aquifer option value	L

- Reductions in the possibility of legal action against the site owner, resulting in fines and legal costs averted. These can be seen as having value to the property owner but are difficult to quantify or monetize.
- Reduction in aquifer damage and the resulting protection of the aquifer's total economic value (TEV). By preventing further degradation of the aquifer over time or by remediating over time the volume of aquifer already impacted, an increase in the aquifer's TEV is realized. In this case, because the aquifer is not currently used to produce economic benefits, a more accurate valuation would involve the aquifer's option value (the value that society would place on having the option to use the aquifer at some time in the future). However, as stated earlier, the intent of this analysis is to provide a maximum likely estimate of the benefits of remediation to determine a threshold for economic remediation. Thus, treating the aquifer as if it were being used is justified.
- Prevention of further perceived damage to the stream, manifested as a gradual improvement in river quality designation over time (partially monetizable).
- Avoidance of or improvement in perceptions of human and ecological health in the neighboring community, as a result of prevention of non-

compliant HTO discharges to the rivers and treatment facilities in the area (monetizable). As discussed in the risk analysis, the risk of human health effects is considered very low, based on currently observed concentrations of HTO in groundwater and surface water near the site. However, it is clear that the public and the regulators are particularly sensitive to radio-active contamination and that a perception of damage may likely exist within those groups.

15.4.4 Identification and Valuation of Readily Quantifiable Benefits

Of the likely benefits identified previously, only a few are readily quantifiable in monetary terms. The analysis indicates that there are four such benefit categories, if the approach of equating benefits to avoided damages (see Table 15.9). In this case, benefits of remediation accrue if damage to the river and its users is avoided. Benefits of remediation to the site owner accrue due to elimination of legal costs and fines (but not strictly to society as a whole).

For valuation purposes, the perception of damage to the river, its ecosystems, and public health (which is not in reality affected) can be assumed to be reflected by a blight effect on property values in the neighboring community. This approach is supported by recent research by Gawande and Jenkins-Smith, who found substantial negative effects on property values in areas subjected to transshipment of radioactive wastes in the United States.[1] This is estimated very conservatively by assuming that the values of all of the 1000 residential properties in the neighboring town are reduced by 10% (for the base case) or 25% (for the high case) because of the perception of damage and risk resulting from HTO discharge via groundwater.

It is important to reemphasize that it is not implied that this is the case. Rather, in constructing a conservative (worst-case) economic analysis, it is useful to over-estimate purposely the economic value of perceived risks. In this way, the analysis is intentionally biased toward justification of higher expenditure on remediation. Such an approach can be useful if there is a risk that certain outside stakeholders will claim that the analysis has been skewed in favor of the problem holder. In essence, this approach is very similar to that used in conducting quantitative environmental risk assessments. Taking a conservative approach and assuming maximum likely source concentrations of contaminants, low or no attenuation, rapid migration pathways, and the most sensitive receptors, a maximum likely impact is calculated. If this scenario is found to pose an acceptable risk, then there is reasonable probability that receptors will be protected under a normal range of conditions.

A base case is defined, involving a likely conservative valuation of the river, aquifer, and blight categories. The base case assumes a planning horizon of 20 years (one generation, a standard planning horizon in many jurisdictions) and a 2% discount rate. Remediation is assumed to start immediately (time 0). A highly conservative (high) case is also provided, using a higher river valuation and a much higher market rate for water (showing the effect of a higher monetary value being placed on the damaged aquifer) and assuming a 25% blight factor on property values.

TABLE 15.9
Remedial Benefits — Base Case

Benefit Category	One-Time Benefit ($[£])	Annual Damage Avoided ($[£]/Yr)	Planning Horizon (Yrs)	Valuation Method
Prevention of aquifer degradation to the point where it is perceived as being unusable	–	0.18/m³ (0.10) for 5000 m³/day	20	Conservative valuation, based on current market value of water being rendered unusable by presence of HTO, based on groundwater modeling results
Avoidance of prosecution, litigation, and fines (included in financial analysis alone)	–	0.36 million (0.2 million)	20	Estimated annual fines, based on average awards in courts in this jurisdiction, plus costs of associated legal fees
Prevention of river water quality degrading to the point where the next lowest U.K. river category is reached	–	Negligible < 0.01 million (0.01 million)	20	Willingness to pay to avoid further decline of river water quality (estimate taken from literature)
Perception of damage, resulting in blight to the community, reflected in lower property values	31.0 million (17.0 million)	–	One-time benefit realized upon action removing perception of damage	Conservative assumptions: 1000 properties, 10% blight factor on average property value of $310,000 (£170,000)

15.4.5 Identify and Assess Nonquantifiable Benefits

As discussed previously, there are also benefits that cannot be readily monetized. Many such benefits are considered important and, if properly quantified, of significant value. Such benefits need to be included in the overall analysis. Benefits are subjectively assessed based on whether they are expected to increase overall benefits

TABLE 15.10
Nonquantifiable Benefits

Benefit	Private/ External	Likelihood	Relative Benefit Impact	Applicable to Approach	Uncertainty	Comments
Public relations value to problem holder	P	NL	+	S1, S1P, P1, R1, P2PR	Moderate	Very difficult to quantify the positive effect on the problem holder's business.
Reduced liability to problem holder	P	L–VL	++	S1, S1P, P1 R1, P2PR	Low	Liability for the site remains due to other areas of contamination. Provision (if present) can be removed from firm's balance sheet. Not captured by increase in property value.

in general, by a negligible (0), marginal (+), substantial (++), or considerable (+++) amount. This can be accomplished, in part, by gauging the expected contribution of the nonquantifiable benefits against the magnitude of overall quantifiable benefits. Benefits classified as likely (L) or very likely (VL) in Table 15.7 and Table 15.8 are considered. Table 15.10 indicates their relative impact.

The implication of the information in Table 15.10 is that whatever benefits are monetized, there remain other private benefits of remediation that have not been monetized. These nonmonetized benefits cannot be readily quantified at this time but serve to illustrate that the CBA analysis presented subsequently is partial in nature.

15.4.6 BENEFITS SUMMARY

Based on the preceding information, the benefits of remediation for the base case, over 20 years, are provided in Table 15.11. In all cases, remediation is assumed to start immediately, with the agreement of the regulatory bodies. Note that total remedial benefits, if all benefit categories were realized, range from about $35

TABLE 15.11
Benefits Summary

Benefit Category	Sum of Benefits over 20 yrs @ 2% ($ million [£ million]) Base Case	Sum of Benefits over 20 yrs @ 5% ($ million [£ million]) Base Case	Sum of Benefits over 20 yrs @ 2% ($ million [£ million]) High Case
Aquifer amelioration	5.38 (2.99)	4.09 (2.27)	53.87 (29.93)
Value of fines and legal costs averted	6.25 (3.47)	4.84 (2.69)	6.25 (3.47)
River protection	0.31 (0.17)	0.23 (0.13)	0.62 (0.34)
Perception of damage as blight on property	30.6 (17.0)	30.6 (17.0)	76.5 (42.5)
Total benefits[1]	36.29 (20.16)	34.92 (19.40)	130.98 (72.77)

[1] Excluding avoided fines etc. that are included in financial analysis alone.

million in the base case (at 5% discount rate) to about $130 million (at 2% discount rate) for the high case (excluding avoided fines etc. that are transfer payments).

However, the total value of all benefits that can be realized by a remedial objective/approach or a combination of remedial objectives/approaches is a function of which of these benefit categories are actually realized. Not all benefit categories will be realized by each objective or approach, as shown in Table 15.12. For example, an objective of protecting the current receptor (B) for the river (SPR 1), using approach R1 (containment at river's edge), would realize the benefit of river protection and probably all or part of the blight reduction benefit, but the aquifer would not be cleaned up, so the aquifer amelioration benefit would *not* be realized. For this case, the base case 2% discount rate total benefits over 20 years would be estimated at just under $16 million.

15.5 COST ANALYSIS

Costs are calculated for implementing each approach considered in the analysis. A preliminary estimate of least-cost implementation for each approach isdeveloped. These are later compared to the benefits, developed previously, to complete the CBA and select the most appropriate remedial objectives/approaches to satisfy each SPR linkage.

TABLE 15.12
Benefits Summary by Remedial Approach

Benefit Category	Benefits over 20 yrs @ 2% ($ million [£ million]) Base Case	Benefits over 20 yrs @ 5% ($ million [£ million]) Base Case	Benefits over 20 yrs @ 2% ($ million [£ million]) High Case
S1	27.54 (15.30)	27.54 (15.3)	68.85 (38.25)
S1P	22.95 (12.75)	22.95 (12.75)	57.38 (31.88)
P1	13.34 (7.41)	11.97 (6.65)	73.62 (40.90)
P2PR	18.14 (10.08)	17.46 (9.7)	65.50 (36.39)
R1	15.61 (8.67)	15.53 (8.63)	38.86 (21.59)
MNA	3.01 (1.67)	2.29 (1.27)	27.56 (15.31)
S1P+P1	33.23 (18.46)	31.86 (17.7)	123.33 (68.52)
S1P+P2PR	30.38 (16.88)	29.7 (16.5)	96.10 (53.39)

15.5.1 APPROACH S1 — COMPLETE SOURCE REMOVAL (SITE LANDFILL)

This approach is the complete excavation and removal of the source of HTO in groundwater. By removing the source, it is assumed that the HTO plume will gradually attenuate over time, leading to compliance. It is estimated that an area of 3.5 hectares would be removed to an average depth of 10 m, at a unit cost of $450/m³ (£250/m³). This provides a total estimated cost of $160 million (£89 million), which notionally includes all engineering, laboratory, reporting, supervision contracting, and disposal costs. For this analysis, this cost is assumed to be a one-time charge incurred at the beginning of the project lifetime. Soil disposal costs could push this estimate higher if all soil were assumed to be radioactive. Material could be screened and sorted on site before disposal to minimize radioactive volumes. However, this could be technically challenging.

15.5.2 APPROACH S1P — PARTIAL SOURCE REMOVAL (SITE LANDFILL)

The sources of tritium in the tip could exist as small discrete and dispersed masses of material. The actual mass of tritium in the landfill is thought to be quite small. On this basis, excavation of only those areas where tritium exists would result in substantial savings. Notionally, such a program would require the following steps:

- Further investigation to minimize excavation volumes (using a CPT survey and selected groundwater monitoring wells to isolate hotspots). Cost: $1.8 million (£1 million).

- Selected excavation of 25,000 m³ of hotspot material: unit cost $180/m³ (£100/m³·). Cost: $4.5 million (£2.5 million).
- On-site sorting excavated material to further minimize volumes that would go off as radioactive, and disposal off site of radioactive material. Cost: $11.7 million (£6.5 million).

This provides a total all-in cost of approximately $18 million (£10 million).

15.5.3 APPROACH P1 — REMOVAL AND TREATMENT OF GROUNDWATER

This approach involves the control and capture of the current HTO plume in shallow groundwater that emanates from the tip. The cost estimate for application of this approach is based on a pump-and-treat design and requires the following actions, discussed later:

- Estimate flow rate required to capture plume using groundwater modeling
- Estimate concentrations of tritium, hydrocarbons, solvents, iron, etc., at this flow rate
- Estimate cost of any pretreatment
- Estimate cost of tritium treatment, using a patented enrichment process
- Estimate cost of disposal of enriched effluent

15.5.3.1 Groundwater Modeling

A numerical flow model was developed to simulate groundwater flow at the site for the purpose of predicting the volume of water in the gravel aquifer requiring abstraction in order to contain the water flowing through the tritium-contaminated area. Calibration to winter conditions ensured that worst-case conditions were used.

The model shows that, in the most conservative case, three abstraction wells pumping a total of 175 m³/day would be required to achieve containment. The wells would need to be distributed around the western edge of the delineated tritium plume. Three wells pumping at about 32 m³/day each would be required to capture the plume containing tritiated groundwater at concentrations exceeding 50 Bq/l. Total abstraction would be 96 m³/day for the most likely scenario. Using worst-case estimates, the flow rate required would increase to 175 m³/day.

15.5.3.2 Treatment System

Some pretreatment would be required to protect the catalyst used in the proprietary tritium concentration process. The final polishing system would consist of particulate filter, activated carbon, and ion exchange resins to deal with other organic compounds present in the plume. Pretreatment capital costs are estimated at $2.7 million (£1.5 million), with annual operation and maintenance costs of $0.18 million (£0.1 million). Assuming that the input tritium concentration is < 500 Bq/l, a tenfold decrease

TABLE 15.13
Range of Groundwater Treatment Costs for Tritium

Processing Rate (m³/day)	Estimated Capital Cost ($ million [£ million])	Estimated Operating Cost ($ million/yr [£ million/yr])	Power Costs ($ million/yr [£ million/yr])	Pretreatment Cost ($ million/yr [£ million/yr])
96 — likely	23.4 (13.0)	12.1 (6.7)	10.8 (6.0)	2.7 (1.5)
175 — worst case	43.4 (24.1)	13.1 (7.3)	19.8 (11)	3.2 (1.8)

would meet the 50 Bq/l discharge requirement. Cost ranges for tritium treatment are provided in Table 15.13. Three ongoing annual costs are provided:

- Operating costs for the main system
- Operating costs for the pretreatment system
- Overall system power costs

The total present value cost over a 20-year period for the likely case, using the 2% discount rate, is estimated at about $430 million (£240 million), not including costs for final disposal or temporary storage of enriched tritium effluent. For this analysis, these costs have not been estimated, so these treatment costs are lower than expected actual costs.

15.5.4 APPROACH MNA — INSTITUTIONAL MANAGEMENT OPTION

The MNA (monitored natural attenuation) approach essentially involves a detailed and focused monitoring effort of the current groundwater plume and of the concentrations and fluxes of tritium on all identified receptors. The costs associated with this approach include:

- Monitoring, including expanding and improving the monitoring network
- Risk analysis work based on collected data
- Managing the issue with the public and the regulators over the planning horizon

It is assumed that an initial $1.8 million (£1 million) capital upgrade expenditure would be followed by annual monitoring costs of $0.9 million (£0.5 million).

15.5.5 APPROACH P2PR — EXPERIMENTAL RESEARCH APPROACH

This potential approach is based on the premise that new ideas could be encouraged for the management of tritium in groundwater. Some recent research in the U.S. and the U.K. has suggested that phytoremediation techniques could show promise for this type of problem (despite problems with direct evapo-transpiration of HTO through leaves into the atmosphere). For development of a cost estimate, it is assumed that the problem holder would take action to develop a comprehensive research program into cost-effective and innovative remedial techniques for tritium in groundwater, investing as much as $1.8 million (£1 million) each year over the next three years in research and development (R&D), culminating in implementation of a passive or semipassive treatment and containment system, which could be installed and operated for costs much lower than currently available technology. Purely for costing purposes, a capital cost of $7.2 million (£4 million) and O&M costs of $1.35 million (£0.75 million) per year are assumed for implementation of such a system.

Clearly, these costs are hypothetical. However, including such a program in the economic analysis illustrates the potential for developing a comprehensive research program, the aim of which is to develop new, low-cost approaches for tritium in groundwater. In reality, it is recognized that this issue is an international issue for which a national or international research initiative could be developed.

15.5.6 Approach R1 — Collection at the River

This approach involves passive collection of all tritium-contaminated groundwater at the stream edge, as a way of protecting the stream. This could be accomplished through construction of a cofferdam or similar structure at the stream's edge and treating all collected water using the system described earlier for the P1 option. Again, final disposal or storage costs for the enriched water have not been included.

15.5.7 Cost Estimate Summary

Table 15.14 provides the basis for developing preliminary cost estimates for the least-cost implementation of the short-listed approaches. A probability of success is assigned to each approach using the selected least-cost technology. Definitions of the designation are provided in Table 15.15. Table 15.16 classifies each of the proposed approaches.

15.6 COST–BENEFIT ANALYSIS

15.6.1 Results

Preliminary high-level cost and benefit functions have been developed for each approach considered. These are compared in a partial CBA (full costs compared to readily monetizable benefits), qualified by the other factors discussed earlier. Table 15.17 provides the BCRs (benefit–cost ratios, or sum of PV benefits divided by sum of PV costs) and qualifiers for each of the short-listed approaches, and two additional approach combinations, for the base case involving immediate start of remediation and a 2% discount rate. A BCR greater than 1.0 indicates that the benefits of the remedial objective/approach exceed the costs, and thus it is economically worthwhile to achieve that objective. The higher the BCR, the greater the net benefit to society (including the problem holder and the rest of society). The results in Table 15.17

TABLE 15.14
Cost Analysis of Remedial Approaches

Remedial Approach	Remedial Technology for Costing	Capital Cost	Annual Operation Costs	Operation Time (Yrs)	Assumptions
S1	Remove solid tritium from landfill (remove full volume of landfill)	$156 million (£87 million)	None	1	Material is disposed of in secure landfill after screening
S1P	Partial removal of solid tritium from landfill (minimize removal through further investigation)	$18 million (£10 million)	None	1	Further investigation can pinpoint major tritium sources; material is disposed of in secure landfill
P2PR	Research and development into tritiated groundwater; lower-cost methods developed	R&D: $1.8 million/yr (£1 million/yr) for 3 years Implement: $7.2 million (£4 million) in year 5	$1.35 million (£0.75 million)	20	Three-year R&D program, followed by full-scale implementation (if successful)
R1	Collect and treat tritiated water at river discharge point	$3.6 million (£2 million) collection system and $33.3 million (£18.5 million) treatment system	$23.8 million (£13.2 million)	20	Cofferdam or barrier collects plume before discharge to river; treatment and discharge
P1	Remove and treat tritiated groundwater from gravel aquifer between site tip and north ponds	$1.35 million (£0.75 million) pumping system and $33.3 million (£18.5 million) treatment system	$23.8 million (£13.2 million)	20	Pump-and-treat system captures and treats entire plume
MNA	Allow natural attenuation to act on plume	$1.8 million (£1.0 million)	$0.9 million (£0.5 million)	20	Assumes MNA effective and acceptable to regulators

TABLE 15.15
Remedial Technology Category Designations

- *Accepted: High probability of success* — Method has been extensively applied under a variety of conditions, is well documented, and has wide acceptance in the industry. Examples: pump-and-treat, slurry walls, air-stripping for VOC removal from pumped water, etc.

- *Emerging: Moderate probability of success* — Method has been tested at field-scale but is still not fully understood or developed and requires more detailed field validation. Examples: *in situ* air-sparging, funnel-and-gate, natural attenuation, *in situ* bioremediation, etc.

- *Experimental: Low probability of success* — Method is undergoing bench-scale or pilot-scale testing, is not well understood, and is currently the subject of significant study. Examples: semi-passive barriers using Fenton's reagent, phytoremediation, *in situ* surfactant washing, etc.

TABLE 15.16
Probability of Success of Remediation

Remedial Approach	Remedial Technology for Costing	Assumptions	Designation	Probability of Success
S1	Complete excavation and treatment of site tip material	Material is landfilled	Accepted	High
S1P	Partial excavation of site tip material	Material is landfilled	Accepted	High to moderate
P1	Remove and treat tritiated groundwater from gravel aquifer between site tip and north ponds	System must remain operational over long term	Accepted	High to moderate (technically difficult)
P2PR	Phytoremediation to control and treat plume in place	Research provides successful new methods	Experimental	Unknown to low
R1	Collect and treat tritiated water at river discharge point	System must remain operational over long term	Accepted	High
N3MNA	Monitored natural attenuation of groundwater contamination over time, assuming MNA is effective	MNA is effective and acceptable to EA	Emerging	Low to moderate

TABLE 15.17
Approach Selection — Benefit–Cost Ratios for the Base Case (2%)

Remedial Approach	PV Cost ($ Million [£ Million])	PV Benefit ($ Million [£ Million])	Net Present Value ($ Million [£ Million])	BCR	Nonquantifiable Benefit Impact
S1	156.6 (87.0)	27.54 (15.30)	−129.1 (−71.7)	0.18	++
S1P	18.0 (10.0)	22.95 (12.75)	4.95 (2.75)	1.28	++
P1	448.2 (249.0)	13.34 (7.41)	−434.7 (−241.5)	0.03	++
P2PR	27.0 (15.0)	18.14 (10.08)	−8.8 (−4.9)	0.67	++
R1	446.4 (248.0)	15.61 (8.67)	−431.1 (−239.5)	0.03	++
MNA	17.5 (9.7)	3.01 (1.67)	−14.4 (−8.0)	0.17	0
S1P+P1	604.6 (335.9)	33.23 (18.46)	−571.3 (−317.4)	0.05	++
S1P+P2PR	45.0 (25.0)	30.38 (16.88)	−14.6 (−8.1)	0.68	++

are represented graphically in Figure 15.1, which shows the BCR for each of the eight approaches examined in the analysis, in the order of lowest- to highest-cost approach.

15.6.2 DISCUSSION

Examination of the CBA results in Table 15.18 and Figure 15.1 leads to the following observations. The first is that the benefit–cost ratio is maximized when approach S1P (partial excavation of contaminated sediments) is used (BCR = 1.28). If a tritium clean-up target of the current WHO drinking-water levels were agreed upon, this approach might be capable of achieving outright objective B (reduce impact to current receptor) for both of the SPR risk linkages. Attempting to achieve a target of background 50 Bq/l (objective C), however, would mean that this approach could not be used to deal with any of the SPR linkages. However, note that the net benefit of S1P is relatively small, at $5 million (£2.8 million), compared to the substantial negative net benefit values for many of the other approaches (negative $130 million [£72 million] for S1 and negative $436 million [£242 million] for P1, for example). This illustrates that BCR, although a useful index, does not reflect the magnitude of the actual costs or benefits, and so alone does not convey some of the key information decision makers will require.

   Figure 15.1, which lists approaches in order of increasing cost, shows clearly that just because a remedial approach is more expensive does not mean that it will provide more benefit. Also, note that the least costly remedial approach is not the most economic. These trends are seen frequently when conducting CBA for remedial decision making for groundwater and contaminated sites in general. In most circumstances, an economic optimum lies at some point between the cheapest and most expensive alternatives.

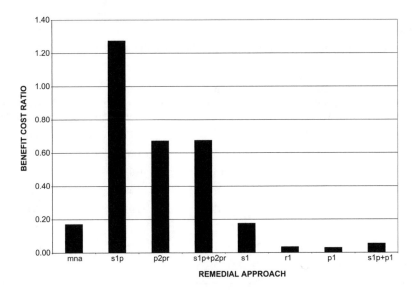

FIGURE 15.1 Benefit–cost ratio — base case (discount rate of 2%).

Combining this limited source removal with a program of active hydraulic pumping (S1P+P1) satisfies objective C but with a much lower BCR (0.05), reflecting the very high costs of water treatment. In fact, it is clear that all of the approaches capable of satisfying objective C (which relates directly to achieving a background target of 50 Bq/l HTO) have very low BCR (all below 0.05) and very high 20-year PV costs (all above $430 million [£240 million]). From a purely economic perspective, this result clearly indicates that attempting to achieve background groundwater quality or HTO at the site is not in the best interests of society as a whole. Other objectives provide much greater value to all stakeholders for the investment required.

Two other approaches exhibit BCR above 0.5 for the base case. Although the cut-off point (or point of indifference) is a BCR of 1 (benefits are equal to costs), it is worth looking at these approaches as well. The research approach (P2PR), involving a concerted effort at funding and developing new and cost-effective techniques — perhaps as part of an international effort — has a BCR of 0.67, which, given the inherent uncertainties, is approaching economic feasibility compared to most other available options (especially considering its "substantial (++)" rating for nonquantifiable benefits). Combined with S1P (partial removal of sources in the tip), the two approaches together would have a BCR of about 0.7. In both cases, we can expect that other nonmonetized benefits could possibly lift the actual BCR close to unity. In addition, P2PR, both alone and combined with S1P, is capable of managing all of the SPR linkages. This analysis is akin to the recommendations about combining monetary and nonmonetary benefit assessments in Chapter 5.

TABLE 15.18
Approach Selection — Benefit–Cost Ratios for the Base Case (5%)

Remedial Approach	PV Cost ($ Million [£ Million])	PV Benefit ($ Million [£ Million])	BCR Base Case 5% Discount Rate	BCR Base Case 2% Discount Rate
S1	156.6 (87)	27.54 (15.3)	0.17	0.18
S1P	18.0 (10)	22.95 (12.75)	1.28	1.28
P1	331 (184)	11.97 (6.65)	0.03	0.04
P2PR	25.9 (14.38)	17.46 (9.7)	0.67	0.67
R1	334 (185.5)	15.53 (8.63)	0.03	0.04
MNA	13.1 (7.25)	2.29 (1.27)	0.17	0.17
S1P+P1	350 (194)	31.86 (17.7)	0.05	0.09
S1P+P2PR	43.9 (24.38)	29.7 (16.5)	0.68	0.68

15.6.3 SENSITIVITY ANALYSIS

Several of the key input parameters in the analysis are subject to uncertainty. It is instructive to examine how variation of these parameters may affect the overall result of the analysis and, ultimately, the choice of remedial objectives and approaches. In particular, variations in the following parameters are examined:

- Discount rate (typically, a lower discount rate places a higher value on the conservation of resources in the future and less emphasis on the need to realize positive benefits quickly)
- Timing of remediation
- Valuation of groundwater, river impacts, and blight

15.6.3.1 Discount Rate

To examine the impact of a higher discount rate on the analysis, a discount rate of 5% has been used. Both costs and benefit flows were discounted, as with the 2% base case.

Table 15.18 shows the PV costs, benefits, and BCR for each approach, for the base case using a 5% discount rate. Base case 2% discount rate BCRs are also provided for comparison. The overall effect on BCR is relatively insignificant, with BCR generally declining slightly, especially for approaches that are generally more economically feasible (BCR closer to unity). At 2%, the present value of ongoing operation and maintenance costs (for approaches where these are required) are higher than for the 5% base case.

These data are presented graphically in Figure 15.2, with approaches presented in order of increasing cost (least cost on the left, highest cost on the right). The effect of the change in discount rate on the BCR can be seen, but the overall ranking of approaches by BCR does not change.

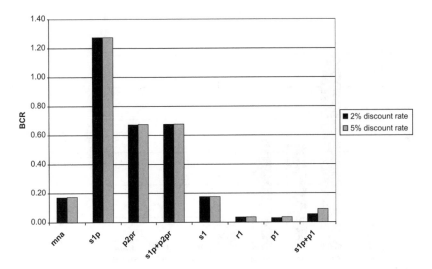

FIGURE 15.2 Benefit–cost ratios with two different discount rates (2% and 5%).

TABLE 15.19
Effect of Delay on Start of Remediation (Base Case at 2%)

Remedial Approach	PV Cost (5-Year Delay) ($ Million [£ Million])	PV Benefit (5-Year Delay) ($ Million [£ Million])	BCR Base Case 2% Discount Rate 5-Year Delay	BCR Base Case 2% Discount Rate Immediate Start
S1	139 (77.43)	24.9 (13.9)	0.18	0.18
S1P	16.0 (8.9)	20.9 (11.6)	1.30	1.28
P1	307.8 (171)	9.7 (5.4)	0.03	0.04
P2PR	23.8 (13.2)	15.3 (8.5)	0.64	0.67
R1	309.6 (172)	14.0 (7.78)	0.05	0.04
MNA	12.1 (6.7)	1.5 (0.82)	0.12	0.17
S1P+P1	324 (180)	27.7 (15.4)	0.09	0.09
S1P+P2PR	39.8 (22.1)	26.3 (14.6)	0.66	0.68

15.6.3.2 Timing of Remediation

The analysis has been conducted considering a five-year delay in the start of remediation, for the base case, at 2% discount rate. The results are shown in Table 15.19. Comparison of BCRs for an immediate start and a five-year delay is provided in Figure 15.3, again with approaches presented in order of increasing cost, left to right.

Comparing Table 15.17 with Table 15.19 shows that the result of deferral of action is considerably lower PV benefits and costs, as expected. At higher discount rates, the effect would be even greater. However, the overall impact on BCR is relatively small and does not affect the overall ranking of the approaches or their ability to manage the SPR risk linkages at the site.

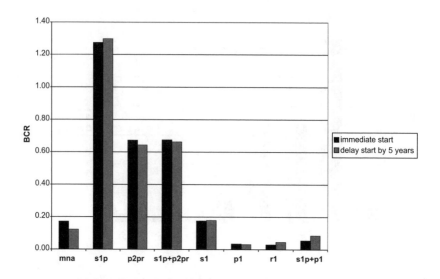

FIGURE 15.3 Benefit–cost ratios — different start times (immediate and delay of 5 years).

15.6.3.3 High Benefits Valuation Scenario

One of the major limitations of the benefits analysis conducted for the example site is the lack of reliable and robust studies on aquifer valuation. In the benefits analysis for this site, an approximate valuation of the value of groundwater was developed using a current average sale price of water to domestic users in the site area, divided by a factor representing the costs of abstraction, treatment, and delivery. This value was then used to estimate the average replacement cost for water rendered unusable by the presence of contamination in the aquifer. A figure of $1.8/m³ (£1/m³) was used for the value of groundwater (a tenfold increase over the base case). As discussed earlier, even this approach is highly conservative; in other words, it overvalues the aquifer by assuming the water has a current use value, which in reality it does not. However, we might consider that the aquifer has some option value that is not reflected in market price.

In the same way, to strive for an analysis scenario that is highly conservative, higher valuations have been placed on river value (double the base case) and on blight (by elevating the blight factor to 25% of current property value, as a one-time benefit, realized upon acceptable management of the groundwater issue to the satisfaction of the regulator).

15.6.3.4 Results

Table 15.20 shows the costs and benefits of each of the eight approaches considered, with the increased value of benefits (property, aquifer, and river). The BCR for the high-value case is provided along with the BCR for the base case. Figure 15.4 shows the costs of each approach, from least to most expensive, along with the base case benefits and higher valued benefits.

TABLE 15.20
High Valuation Case CBA (at 2%)

Remedial Approach	PV Cost ($ Million [£ Million])	PV Benefit (High Case) ($ Million [£ Million])	BCR High Case 2% Discount Rate Immediate Start	BCR Base Case 2% Discount Rate Immediate Start
S1	156.6 (87.0)	68.85 (38.25)	0.44	0.18
S1P	18.0 (10.0)	57.38 (31.88)	3.19	1.28
P1	448.2 (249.0)	73.62 (40.90)	0.16	0.04
P2PR	27.0 (15.0)	65.50 (36.39)	2.43	0.67
R1	446.4 (248.0)	38.86 (21 59)	0.09	0.04
MNA	17.5 (9.7)	27.56 (15.31)	1.56	0.17
S1P+P1	604.6 (335.9)	123.34 (68.52)	0.20	0.09
S1P+P2PR	45.0 (25.0)	96.10 (53.39)	2.14	0.68

As expected, all approaches show increases in net benefit, characterized by increases in BCR. However, the overall conclusions of the analysis, discussed for the base case, remain substantially unchanged. BCR is maximized by implementing a limited source removal approach (S1P). A focused research program leading to control of the plume moves from negative to positive net benefits (BCR > 1). Both remain capable of dealing with all of the identified SPR linkages.

Figure 15.4 illustrates the effects of increasing benefit values on the overall economics of the remedial approaches. Even adding a highly conservative valuation, which in effect justifies higher remedial expenditures, does not change the fundamental results. Remedial approaches involving an objective of 50 Bq/l in groundwater (background) are not economically justified, even if groundwater is valued 10 times above its current market level and significant blight factors are put on all properties in the area. The reason is that achieving such a standard at this site, *with currently available technology*, is tremendously expensive. Again, the economic analysis shows that society as a whole is best served by attempting to remove hotspots of tritium contamination from the landfill and by engaging in a concerted research effort to develop new and more cost-effective remedial techniques for tritium in groundwater.

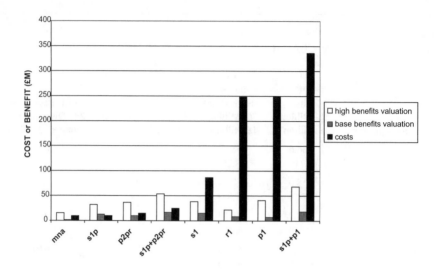

FIGURE 15.4 Costs and benefits. Cost estimates are shown by black bars.

15.7 UNCERTAINTY AND LIMITATIONS

The analysis is subject to a number of inherent limitations and uncertainty, largely the result of assumptions and the often subjective nature of selections and appraisals that must be made. In all such analyses, the results necessarily depend on the expert input of the user in determining which remedial approaches are most appropriate for any given scenario and which risks are most significant and likely. In reality, these are the same limitations inherent in most, if not all, decision-making processes for contaminated sites and groundwater: they depend heavily on the expertise and experience of the user and, in some cases, on the perspective of the user or stakeholder. As such, this methodology is seen as a tool for negotiation between stakeholders, each of whom will tend to value various resources and potential risks slightly differently. Specific limitations of the analysis are:

- Only four main benefit categories are monetized. The effects of other benefits, not readily monetized, are described in a qualitative way. In all of the preceding discussions, it is clear that the benefits of a particular action are at least the monetized value and, in most instances, appreciably above. Refining benefits estimates further than the level presented in this example is possible but in practice unlikely, because doing so would require considerable effort and expense in its own right.
- Value of groundwater in most parts of the world is still poorly understood, as discussed earlier. A greater understanding of this key issue is urgently required if economically optimal decisions involving groundwater remediation are to be made in future.

- Predictions of the success of aquifer remediation in response to a remedial program are extremely uncertain. In the example, various levels of aquifer amelioration were assessed to provide a sensitivity analysis on the effects on CBA. The greater the value placed on groundwater, the greater the impact of uncertainty in predicting aquifer amelioration on CBA.
- Despite using a relatively rigorous series of steps and analysis matrices, final selection of remedial approaches for each objective still requires a significant degree of expertise.
- Costs of remediation are considered to be stable over time, when assessing the impacts of deferral of start time. In reality, technology may change over time, as may regulations and law. Indeed, the value of groundwater as a resource may also change in the future, particularly if the effects of global climate change are considered.
- A 20-year planning horizon was used throughout. Many benefits, such as groundwater improvement, will extend over considerably longer periods of time. Especially at lower discount rates and if groundwater is more highly valued, longer planning horizons may be justified. For radioactive compounds with longer half-lives, this will certainly be the case. However, limitations and uncertainties also exist when longer planning horizons are used. Changes in the value of commodities and resources may fluctuate considerably and cannot reasonably be predicted in many instances. Regulations, law, and technology may also change, with possibly profound effects on costs and benefits. Under these circumstances, a 20-year planning horizon is considered appropriate for the analysis.
- To this point, costs for final disposal of treated water have not been included in the cost side of the analysis. However, as can be seen from the very low BCR values for approaches using currently available treatment methods, adding these costs will not materially affect the results of the analysis but will only reinforce the conclusions drawn.

15.8 SUMMARY

Two source–pathway–receptor (SPR) risk linkages were considered:

- Tritium in groundwater from the landfill discharging to the stream
- Tritium from the landfill impacting the aquifer itself and potential future users of the aquifer

Based on available information and the currently published drinking-water limits for HTO, risks to human health were not considered in the analysis.

Eight remedial approaches and approach combinations were selected for detailed economic analysis. These included:

- Full excavation and disposal of the tip
- Partial excavation of hotspots of tritium within the tip

- Hydraulic pumping of the aquifer to contain the current tritium plume in groundwater, with on-site treatment of contaminated water using a patented concentration system
- Funding of a comprehensive research and development program into innovative, cost-effective treatment methods for tritium in groundwater

Each monetizable benefit is valued using an approach that will provide a larger value than would normally be considered likely. In addition, a qualitative examination of some likely nonmonetizable benefits is included. Thus, in the cost–benefit analysis, likely costs are compared with conservatively high benefits. What results is, in effect, a worst-case economic analysis for the problem holder.

Capturing and treating the tritium plume by pumping the aquifer is very expensive (present value $450 million [£250 million] over 20 years), and proportionately the benefits are very small. Even under the most optimistic benefits valuation scenario (high case), the BCR for this remedial approach is only 0.16 (negative net benefit of over $360 million [£200 million]). In fact, all of the approaches capable of satisfying the background of 50 Bq/l have a very low BCR (all below 1) and very high PV 20-year costs (all above $430 million [£240 million]). From a purely economic perspective, this result indicates that applying this standard is not in the best interests of society as a whole. Other remedial objectives provide much greater value to all stakeholders for the investment required.

For the base case at a 2% discount rate, the BCR is maximized when a remedial approach of partial excavation of the landfill (removing and disposing of tritium hotspots) is used (BCR = 1.28). This represents the highest ratio of benefits to costs. However, this approach could not achieve the remedial objective of restoring the aquifer to background HTO levels. If, however, the current WHO drinking-water levels for HTO were chosen as the target standard, this approach would likely be capable of reducing the impacts to current receptors for both of the SPR linkages.

CBA suggests that the most economically justifiable remedial approaches are, in descending order:

- Limited excavation of hotspots in tip, with continued monitoring
- A focused research program leading to cost-effective implementation of groundwater remediation
- Hotspot removal combined with research into dealing with residual aquifer contamination in innovative and cost-effective ways
- Continued monitoring under the current discharge regime

Under the most likely case, only the first approach yields positive net benefits. If unrealistically high benefits valuations are used, both the first and second approaches are predicted to yield positive net benefits. When nonquantifiable benefits are included (notionally), the first three options will tend to be more economic.

REFERENCE

1. Gawande, K. and Jenkins-Smith, H., Nuclear waste transport and residential property values: estimating effects of perceived risks, *J. Env. Econ. Manage.*, 42, 207, 2001.

16 Example Problem and Solution

16.1 INTRODUCTION

This example problem contains elements of a number of real sites from different parts of the world but is nonetheless fictitious. The intent is to demonstrate the key concepts involved in setting up an economic analysis, including formulating remedial objective alternatives, assessing constraints, comparing stakeholder viewpoints, coping with uncertainty, and examining sensitivity analysis.

16.2 SITE BACKGROUND

16.2.1 SETTING AND HISTORY

The Cajun Chemical Company (3C) owns a 10 acre site within the town of Houellebecq. The site was home to a chemical works from the early 1950s until its closure and demolition in 1981. The site manufactured and used a variety of solvents and other organic compounds, including PCBs and TCE. The site has been closed and fenced off since 1981. At the time of demolition, all above-ground structures and buried tanks were removed and a subsurface investigation was conducted. This investigation concluded that some residual concentrations of solvents were present in the shallow soil, but shallow groundwater was relatively unaffected.

The site is located atop a small hill, on the west side of the main road through town. To the south on both sides of the highway is relatively high-value residential housing. There are 100 homes within a 1-mile radius of the site. To the east of the highway, about 2 miles from the site, is a large park containing playing fields and woods, and farther east is a wetland. The wetland is a well-known bird habitat. Farmland lies beyond the wetland to the east. In this area, farmers are using groundwater from the regional fractured limestone aquifer for irrigation of crops. To the north of the site, on both sides of the highway, are commercial and light industrial businesses.

The geology of the area is composed of unconsolidated surficial sediments (layered sands, gravels, and silts) varying from 10 to 30 feet thick, overlying and in hydraulic connection with a regional fractured limestone aquifer more than 200 feet thick. Regional geological studies characterize the limestone as intensely fractured in its upper 50 feet, becoming less fractured with depth. The matrix is relatively well cemented throughout, with fracture permeability dominating.

16.2.2 CONTAMINATION

Recently, under pressure from the local community and the regulator and after being approached by a large real estate development company that has expressed interest in purchasing the site, 3C conducted a review of site conditions. This led to the decision to undertake an updated site investigation.

The results of the new investigation revealed a much more complex situation than had been thought to exist, based on earlier information. DNAPL liquids were found to be present in the sediments at several locations, and DNAPL was also found to have penetrated at least 50 feet into the fractured limestone aquifer. No DNAPL was found off site. Total NAPL volume was estimated as approximately 500,000 gallons (about 1.9 million liters). The volume of sediment requiring treatment was estimated as about 20,000 m³.

More detailed groundwater monitoring revealed a plume of benzene, xylene, TCE, and other chlorinated organics in groundwater in the sediments extending down-gradient to the east, actually reaching and discharging into the wetland. A preliminary ecological survey indicated that localized damage to the wetland ecosystem has occurred in patches.

Lower dissolved-phase concentrations (but still exceeding MCLs) of the same compounds were also detected in the limestone aquifer. The leading edge of the bedrock plume was eventually determined to lie 3 miles east of the site. It was determined that the plume had not yet reached the point where it would be captured by the nearest of the farm wells, but modeling predicts that this might occur in as little as 5 years without some form of intervention. Under present conditions, farmland totaling 40 hectares could be affected if the plume reached the nearest farm well's capture zone.

16.2.3 RISK ASSESSMENT (BASE CASE)

Results of the risk assessment show the following:

- Risk 1 — The ecology of the wetland is being adversely affected at and near the main discharge points. Concentrations of COC (compounds of concern) at the wetland discharge points are not sufficiently elevated to cause unacceptable human health risks for visitors to the wetland.
- Risk 2 — On-site concentrations in soil are not fit for residential or commercial end-use.
- Risk 3 — Groundwater at farms to the east is not yet affected but is predicted to be affected in the future under present conditions (pumping rates and predicted migration rates). Expected concentration would exceed drinking-water MCLs, and levels are predicted to create crop damage if used for irrigation.

Vapor generation from dissolved-phase plumes in the sediments is not predicted to pose unacceptable risks to households east of the site.

16.2.4 STAKEHOLDERS

Those with a stake in the site's future are:

- Problem holder, 3C — The company now realizes it has a problem on its hands. Management has decided they would like to sell the site and reduce their liability. The team responsible for the site has been told that they have a budget of $10 million to deal with the problem. Greater expenditure could jeopardize the firm's status as a going concern.
- Houellebecq Community Association — The local residents are concerned about the effects of the contamination on their property values and their use of the park and wetland area. They are anxious also that the site be redeveloped, preferably for residential use. They feel this will add to the value of their property.
- Big Atom Development Co. — Would like to acquire the site for redevelopment. This company has approached 3C and said it would take the site and its liability from 3C, based on a payment from 3C of $2 million. 3C has not yet responded to this offer.
- Save Our Wetlands — The local action group concerned with wetland preservation wants complete protection of the wetlands and restoration of damaged areas.
- Regulators — Represent the public's interest and want the law enforced.
- Local government — Would like to see the site redeveloped because this would increase the town's tax base. Therefore, local government is putting pressure on the regulator to help accommodate a solution.

16.3 PROBLEM FORMULATION AND DATA

16.3.1 THE PROBLEM

Based on the framework presented in Part III and following the process of the case studies in the previous chapters of Part IV, you should consider the issues and answer the following questions when formulating the problem:

- What are the possible remedial objective options for dealing with identified risks?
- What feasible remedial approaches would appear on a short-list to achieve each objective, and which technologies would likely be used?
- What might be the external costs of these remedial approaches or technologies?
- What additional technical data may be required?
- What are the interrelationships of these approaches? (Can approaches achieve multiple objectives?)
- What constraints (temporal, physical, financial) exist, and how will they affect decision making?

TABLE 16.1
Indicative Remedial Costs

Approach	Technology	Capital Costs ($ M) (in Year 0)	O&M Costs ($ M per Year)	Present Value of Costs (20 Yrs, 3.5%)
Treat soil to residential	Excavate and thermal treat 30,000 m³	3.6	–	3.6
Treat soil to commercial	Excavate and thermal treat 20,000 m³	2.4	–	2.4
Control of shallow groundwater plume	Pump-and-treat, interception wells	0.7	0.2	3.3
	Sparging barrier	0.5	0.1	1.8
Control of deep groundwater plume	Pump-and-treat interception wells	1.2	0.3	5.2
Remove NAPL in fractured bedrock	In situ S/SEAR (experimental)	4.5	0.4	7.4
Prevent exposure of groundwater users	Build pipeline to connect to municipal supply	5.5	–	5.5
	Compensation for pumping at farm wells	–	0.2	2.6
	Buy farmland	4.0	–	4.0
Combinations of the above				

- What are the likely major benefits accruing from each remedial objective or approach?
- What valuation information is needed to complete an economic analysis of options?
- What are the different stakeholders' views on the problem and how can these be compared?
- What variables might be worth considering in a sensitivity analysis?

16.3.2 INDICATIVE REMEDIAL COSTS

Table 16.1 presents indicative remedial costs for six remediation approaches that could be used for this problem. Both capital and operating and maintenance (O&M) costs are given in total and per year, respectively. The last column in the table shows the present value of costs over 20 years with a discount rate of 3.5%. This is to illustrate the base case calculations and different time periods and discount rates that could be used for alternative runs of the cost–benefit analysis (as part of

TABLE 16.2
Indicative Unit Benefit Estimates — Base Case

Benefit	Valuation Basis	$ M — One Off Benefit	$ M — Per Year Benefit	Present Value of Benefits (20 Yrs, 3.5%)
Value of site for commercial land undeveloped	Market	4	–	4
Value of residential land undeveloped	Market	3	–	3
Blight reduction in community	5% on 100 homes, average value $0.2 million	1	–	1
Value of lost agricultural production from halt in pumping from farm wells	Market value of lost crop production	–	0.2	3.8
Economic value of wetland	WTP estimate from literature	–	0.06	0.8
Value of increased tax revenue from 50 additional residences	Data from local authority	–	0.25	3.3

sensitivity analysis). Note that capital costs are assumed to occur in year 0 (the first full year of the investment) and be complete within that year.

In addition to implementing each approach on its own, you may also want to combine some of the approaches, with resulting changes to the costs and remediation effectiveness. For simplicity, assume that there are no changes to capital and O&M costs, so that the total cost of two remedial approaches combined is the sum of the costs of the individual approaches.

16.3.3 INDICATIVE BENEFIT ESTIMATES

Table 16.2 presents indicative remedial benefit estimates, including both private and public benefits. Private benefits include the value of the site development for commercial and residential use. Wider economic (or public) benefits include the value of farmland, value of lost crop yield, economic value of wetland, and tax revenue to the local authority.

To help with the assessment of benefits over time, two cases are illustrated. The first one, presented in Table 16.3 is the "lower land value" case; the second, presented in Table 16.4, is the "higher aquifer value" case.

TABLE 16.3
Indicative Benefit Values — Lower Land Value Case

Benefit	Valuation Basis	$ M — One Off Benefit	$ M — Per Year Benefit	Present Value of Benefits (20 Yrs, 3.5%)
Value of site for commercial land undeveloped	Market	1.5	–	1.5
Value of residential land undeveloped	Market	1	–	1
Farmland value	Market	0.5	–	0.5
Blight reduction in community	5% on 100 homes, average value $0.2 million	1	–	1
Value of lost agricultural production from halt in pumping from farm wells	Market value of lost crop production	–	0.2	3.8
Economic value of wetland	WTP estimate from literature	–	0.06	0.8
Value of increased tax revenue from 50 additional residences	Data from local authority	–	0.25	3.3

In all benefits tables, the last column shows the present value of benefits, calculated using a 20-year time horizon and 3.5% discount rate — the same parameters as those used for cost calculations. For simplicity, it is assumed that land transactions also happen in year 0.

16.3.4 QUESTIONS

With the information given previously and the framework of cost–benefit analysis presented in this book, we can answer the following questions, among others:

- What should 3C do, if all costs and benefits are considered?
- What should 3C do, if external costs and benefits are not considered?
- Should BAD buy the site? If so, what should they do?
- What should the farmers do?

16.4 SOLUTION

Based on the preceding information, we illustrate the cost–benefit analysis that can be undertaken for seven possible remedial approaches under the three scenarios:

TABLE 16.4
Benefit Values — High Aquifer Value Case

Benefit	Valuation Basis	$ M — One Off Benefit	$ M — Per Year Benefit	Present Value of Benefits (20 Yrs, 3.5%)
Value of site for commercial land undeveloped	Market	4	–	4
Value of residential land undeveloped	Market	3	–	3
Farmland value	Market	1	–	1
Blight reduction in community	5% on 100 homes, average value $0.2 million	1	–	1
Value of lost water production from halt in pumping from PS wells	Market value of lost water production	–	1	13.2
Economic value of wetland	WTP estimate from literature	–	0.06	0.8
Value of increased tax revenue from 50 additional residences	Data from local authority	–	0.25	3.3

- Base case
- Lower land value
- Higher aquifer value as used for irrigation (and valued by value of crop yield)

The remedial approaches are coded to indicate where on the source (S)–pathway (P)–receptor (R) chain each addresses the contamination problem. Other options and other scenarios (for sensitivity analysis) are, of course, possible. Table 16.5, Table 16.6, and Table 16.7 show the calculations of the seven approaches and three scenarios covered here, respectively. A zero-benefit in the tables indicates that the remediation approach of concern does not address the damage to that receptor or does not allow that particular type of redevelopment of the site. All present value (PV) estimates are for 20 years using a 3.5% discount rate, as before.

In the base case shown in Table 16.5, both internal and external costs and benefits are included in the net present value (NPV) and benefit–cost ratio (BCR) calculations. Thus, this is a complete economic analysis from the point of view of the whole society answering the first question in Section 16.3.4. The value of increased tax revenue does not enter the economic analysis as taxes are transfer payments. Note that to keep the example relatively simple, external costs of remedial approaches (see Chapter 8 for description) are not included in the analysis.

TABLE 16.5
Cost–Benefit Analysis for Selected Approaches — Base Case ($ Million)

| Code | Remediation Approach | Remediation Cost ($M) | | | | Remediation Benefits (PV) ($M) | | | | | | |
| | | | | | | Internal | External | | | | | |
		Capex	Opex	Op time (years)	Total PV of costs	Land sale (private)	Blight for neighboring properties	Wetland protection	Value of affected crop yield	Total PV of benefits	NPV= PV Costs – PV Benefits	BCR= PV Benefits/PV Costs (Ratio)
S1	Remediate soil to residential development quality	3.6		1	3.6	3	1	0.4	0	4.4	0.8	1.2
S2	Remediate soil to commercial development quality	2.4		1	2.4	4	1	0.4	0	5.4	3.0	2.2
S3	Remove NAPL from bedrock	2.2	0.4	10	5.1	0	0	0	1.2	1.2	–4.0	0.2
P1	Containment in surficials	0.7	0.2	20	3.3	0	0	0.8	0	0.8	–2.5	0.2
P2	Containment in bedrock	1.2	0.3	20	5.2	0	0	0	2.8	2.8	–2.4	0.5
R1	Treat farm water	0.4	0.1	20	1.7	0	0	0	2.8	2.8	1.1	1.6

Note: Due to rounding, NPV and BCR may not correspond to total cost and benefit columns.

TABLE 16.6
Cost–Benefit Analysis for Selected Approaches — Lower Land Value ($ Million)

| | | Remediation Cost ($M) | | | | Remediation Benefits (PV) ($M) | | | | | | |
| | | | | | | Internal | External | | | | | |
Code	Remediation Approach	Capex	Opex	Op time (years)	Total PV of costs	Land sale (private)	Blight for neighboring properties	Wetland protection	Value of affected crop yield	Total PV of benefits	NPV= PV Costs – PV Benefits	BCR= PV Benefits/PV Costs (Ratio)
S1	Remediate soil to residential development quality	3.6	0.4	1	3.6	1	0.5	0.4	0	1.9	−1.7	0.5
S2	Remediate soil to commercial development quality	2.4		1	2.4	4	0.5	0.4	0	2.4	0	1
S3	Remove NAPL from bedrock	2.2	0.4	10	5.1	0	0	0	1.2	1.2	−4.0	0.2
P1	Containment in surficials	0.7	0.2	20	3.3	0	0	0.8	0	0.8	−2.5	0.2
P2	Containment in bedrock	1.2	0.3	20	5.2	0	0	0	2.8	2.8	−2.4	0.5
R1	Treat farm water	0.4	0.1	20	1.7	0	0	0	2.8	2.8	1.1	1.6

Note: Due to rounding, NPV and BCR may not correspond to total cost and benefit columns.

TABLE 16.7
Cost–Benefit Analysis for Selected Approaches — Higher Aquifer Value ($ Million)

Code	Remediation Approach	Remediation Cost ($M)				Remediation Benefits (PV) ($M)					NPV= PV Costs – PV Benefits	BCR= PV Benefits/PV Costs (Ratio)
						Internal	External					
		Capex	Opex	Op time (years)	Total PV of costs	Land sale (private)	Blight for neighboring properties	Wetland protection	Value of aquifer	Total PV of benefits		
S1	Remediate soil to residential development quality	3.6		1	3.6	3	1	0.4	0	4.4	0.8	1.2
S2	Remediate soil to commercial development quality	2.4		1	2.4	4	1	0.4	0	5.4	3.0	2.2
S3	Remove NAPL from bedrock	2.2	0.4	10	5.1	0	0	0	5.9	5.9	0.8	1.1
P1	Containment in surficials	0.7	0.2	20	3.3	0	0	0.8	0	0.8	-2.5	0.2
P2	Containment in bedrock	1.2	0.3	20	5.2	0	0	0	13.2	13.2	8	2.5
R2	Treat water	4.5	0.4	20	9.8	0	0	0	13.2	13.2	3.4	1.3

Note: Due to rounding, NPV and BCR may not correspond to total cost and benefit columns.

Table 16.5 shows that remedial approaches S3, P1, and P2 fail the NPV and BCR tests simply because they do not generate sufficiently large benefits compared to their costs, even when wider benefits to the whole society are considered. Recall that a negative NPV and a BCR of less than 1 show that costs exceed benefits. On the other hand, remedial approaches S1, S2, and R1 pass these two tests. The approach with the highest positive NPV, which also happens to be the one with the highest BCR, is approach S2. CBA would recommend this approach in this case.

CBA is only one tool of decision-making analysis, and other factors outside the scope of CBA or costs and benefits that cannot be quantified in monetary terms could also play a role in decision making. For example, approach R1 looks better than approach S1 on the basis of NPV and BCR alone. However, R1 can only address the contamination problem as experienced by the farmers and does not benefit the problem holder, the neighboring community, or the wetland. S1, on the other hand, generates benefits in these three categories but does not protect the farmers. This is where CBA can be used as a negotiating tool among the problem holder, the local community, and the farmers, allowing comparison of the combinations of approaches and the main technical options. For example, S1 plus the purchase of farm land could be compared with R1, and so on.

Table 16.6 undertakes the same analysis for the same remedial approaches but uses the values for the lower land value case. The results are somewhat different from those presented in Table 16.5. All approaches perform worse under this scenario than for the base case. In fact, this time R1 would be recommended purely on the basis of the CBA.

Table 16.7 shows the results of the CBA for the same six remedial approaches using high aquifer value as used by agriculture and measured by the value of crop yield. Table 16.7 shows that the recommended approach here is P2, which provides benefits in terms of addressing the benefits to wetlands and the aquifer, even though it does not allow for site redevelopment.

The other questions asked in Section 16.3.4 can also be answered by analyzing Table 16.5 through Table 16.7. For example, the second question (what should 3C do if only its costs and benefits are considered?) can be answered by comparing the cost of remedial options to private benefits alone. This comparison shows that, in the base case, the recommended approach would be S2 (NPV of $1.6 million and BCR of 1.5); in the lower land value scenario, no remediation would be worthwhile to 3C, and in the higher aquifer value scenario, 3C would still go for S2 because the aquifer value does not accrue to the firm.

In fact, it seems that 3C could clean up the site for an amount lower than its allocated budget and so could consider spending some of the remaining funds to generate some external benefits for the community, at the same time improving the company's image. Interestingly, under the assumptions that could be described as best for 3C (the base case), the net benefit is only $1.6 million, whereas the prospective buyer, BAD, is offering 3C $2 million for the site with all its liabilities. From a purely commercial point of view, it would be better for 3C to sell the land (with a net benefit of $0.4 million compared to the case where 3C undertakes the remediation itself). However, other factors, such as improved company image, may suggest that 3C should undertake the clean-up itself. The decision, as always, lies with the relevant decision makers.

Whether or not BAD should buy the site depends on the comparison of costs of land remediation, purchase, and development with the potential revenue from the sale of the land. Similarly to the 3C's situation and considerations, BAD would go ahead with this venture only if the land values are high, as they are in the base case.

Finally, farmers may have to compare the present value of the income from farming with the revenue from the sale of land. This comparison is not shown here and is not the only factor that would play a role in a farmer's decision. Note that the increase in tax revenue for the local authority due to new commercial and residential properties that could be built on the site is not included in the CBA here. These revenues are transfer payments (see Chapter 4 and Chapter 5), so they do not constitute a net change in society's welfare.

Part V

Summary and Conclusions

17 Summary and Conclusions

17.1 SUMMARY

17.1.1 GROUNDWATER RESOURCES IN CONTEXT

All life on Earth depends on water. Groundwater is one of our most precious resources, and one of our most fragile. Of all the water on the planet, only 3% is fresh, and of that the majority is locked away as snow and ice at the poles. Less than a third of the fresh water on earth is actually available to support the ecosphere, and of this more than half is groundwater. Hundreds of millions of people all over the world, in developed and developing nations, depend directly on groundwater for their daily household needs. In some arid parts of the planet, groundwater is literally the sole source of water. Groundwater is also a critical part of the biosphere, feeding rivers, lakes, and wetland systems, sustaining life for untold species, and helping to maintain the earth's threatened biodiversity.

But despite being hidden away under the earth's surface, groundwater is vulnerable. It can be polluted by a wide variety of human activities. Perhaps the biggest concern with underground contamination is that we cannot see it when it is happening, and we do not know where it is going until it gets there. It moves often only a few meters a year, until one day it appears meters or kilometers from the original source.

Tracing part-per-million or even part-per-billion levels of specific pollutants through an aquifer that may be tens or even hundreds of meters below ground is a highly technical and, in many cases an expensive undertaking. Hydrogeologists use many techniques, including drilling and sampling of exploratory wells, to develop a picture of the extent and concentrations of the contaminants. But the subsurface is highly heterogeneous. Usually, the best that scientists can do is to come up with an incomplete view of the likely extent of contamination and some idea of its severity.

Reclaiming polluted aquifers can be expensive, technically difficult, and time consuming. Once a contaminant is introduced into the aquifer, it may take decades or even centuries to flush out completely. Deciding if and when to remediate, and to what degree, can be regarded in the context of alternative environmentally and socially beneficial actions. What else can be done with the money required to restore an aquifer? And if an aquifer must be restored, what is the best way of doing it, which will result in benefits to all those who have an interest in the resource? Under the "polluter pays" principle, increasingly adopted as the fundamental ethical precept for remediation policy, the responsibility for planning, funding, and executing remediation is borne by the polluter. But others also may want a say in what happens, and how. The public, neighbors, other businesses, and the governments that represent them may all have a real stake in the future viability of the resource. Combining and prioritizing these diverse interests into a decision-making process, using a common unit of value, is essential if equitable, practical, and rational decisions are to be made.

Economic analysis can be used to make better decisions about the protection and restoration of one small part of our environment: the water that is found beneath the earth's surface — groundwater.

17.1.2 LEGISLATION

When faced with a decision regarding remediation of a contaminated site or aquifer, a problem holder (private firm, organization, or individual) will conduct its own financial analysis of the project to determine whether to proceed. The anticipated costs of remediation will be compared to the benefits the firm expects to receive, such as increased land value or reduction in corporate liability. This is not strictly an economic analysis, because it considers only the costs and benefits of the problem holder, not of society as a whole. A whole range of other stake-holders may have an interest in the remediation of the site, including neighbors, environmental groups, and owners or custodians of resources that may be impacted by the contamination. It is the role of the regulatory bodies to represent the interests of society as a whole when considering contaminated sites and their remediation. As such, many jurisdictions have recently enacted legislation or guidance that calls for the full costs and benefits of remediation to be assessed as part of the decision-making process. In many places, remedial decision-making guidelines and legis-lation focus on protection or remediation of groundwater and economically valu-able aquifers. Regulations and guidance have been developed dealing with the protection and remediation of groundwater. Different countries and jurisdictions have developed their own procedures and laws, but many share key common elements. Most have promulgated water laws, making it illegal to knowingly pollute usable groundwater resources. Most have adopted risk-based approaches to remediation, and most have made it the responsibility of the polluter to pay for clean-up. In the European Union, and in particular the U.K., the legislation calls for the comparison of costs and benefits, such as through cost–benefit analysis (CBA), to help determine the best way to deal with groundwater contamination.

17.1.3 RISKS FROM GROUNDWATER CONTAMINATION

The risks associated with groundwater contamination can be classified into three categories:

- Risks of damage to groundwater resources themselves (aquifers), and thus to the users of that groundwater (humans, crops, animals)

- Risks of impact to surface water resources, as a result of groundwater's contribution to the resource (via baseflow discharge), and thus to the users of the surface water (humans, crops, animals, ecosystems)
- Risks of impact to receptors as a result of contaminant migration via groundwater (as a risk pathway), including ecosystems, property, natural amenity features, and possibly humans and animals

Because groundwater and the contaminants within it are mobile, impacts may occur at substantial distances from the original source of the contamination. Due to the heterogeneity of geological materials, the patterns and velocities of contaminant movement in groundwater are difficult to predict, and there is significant uncertainty involved in any prediction of future impacts.

The tools of environmental risk assessment are used to identify, qualify, and, if necessary, quantify the risks and impacts to receptors as a result of groundwater contamination. Risks are assessed by considering sources, pathways, and receptors (SPR). Where a receptor is linked to a source of contamination by a viable pathway, a risk may exist. Quantification of damage, or potential damage, to receptors is a key input into estimation of the benefits of remediation.

17.1.4 GROUNDWATER REMEDIATION

In developing a methodology for applying economic analysis to groundwater problems, clear terminology is vital. There exists in the literature today no single set of terms that clearly defines the various stages of remedial design and decision making. Clear distinctions among the different levels at which remedial decisions are made are required if costs and benefits are to be assigned to competing options as part of an economic analysis. For this reason, the following terms are defined:

- Remedial objective — The overall intent of the remediation project. Objectives could include the degree to which groundwater is to be remediated, the protection of specific receptors, or the elimination or reduction of certain unacceptable risks. Remedial objectives are limited in number and are based on receptor protection.
- Remedial approach — The conceptual manner in which the objective is to be reached. Remedial approaches refer specifically to measures that break the source–pathway–receptor (SPR) linkage, by removing part or the entire source of contamination, cutting the pathway, or isolating or removing the receptor.
- Remedial technologies — The specific tools that form the components of the approach. For example, physical containment can be achieved through use of slurry walls, sheet pile walls, or liners, often in conjunction with groundwater pumping and treatment. Source removal can be achieved through excavation and on-site treatment of contaminated soils (by a variety of techniques) or through many available *in situ* techniques. A remedial solution will very often involve the use of several different remedial technologies.

Remedial objectives should be known before detailed design of a remediation program occurs. The choice of a remedial approach is the critical intermediate step, which can be used both as a tool to help set objectives (by considering and comparing various approaches at the conceptual level) and as a guide to the selection of the technological components that will make up the final design. The remedial approach is the level at which comparative economic analysis can most readily be carried out.

The successful remediation of groundwater requires that a number of critical steps be performed before reaching the remedial design stage. The inherent complexities and uncertainties of groundwater contamination mean that implementing groundwater remediation programs can be expensive and time consuming. Most workers in the field recommend following a rational, step-by-step decision-making process. Such an approach should include the following steps:

1. Understand the problems at the site — Through proper site characterization.
2. Assess the risks posed by the problem — The tools of risk assessment are used. These risks can be valued and expressed in monetary terms.
3. Set remedial goals and constraints for the site — Understanding the true and total costs and benefits (to all of society) of remediation is a key consideration.
4. Identify the best practicable remedial approach and technology — Technical and economic analysis allows various possible remedial approaches to be evaluated and compared. Choosing the right approach, and the best technology to implement that approach, requires experience, training, and insight into the wider issues involved. This book presents a detailed framework for using economic analysis of cost and benefits, to select the remedial objective, approach, and technology that will maximize benefits to all of society, including the problem holder.
5. Test and implement the remediation program — Once a remedial approach has been selected, and with it a preferred technology, the technique should be tested at bench-scale (if required) and on a small scale (pilot-scale) under site conditions. Based on the results of pilot testing, the system can be scaled up and implemented at full scale.
6. Monitor results — Assess remedial progress through careful monitoring, and modify as necessary for efficient improvements. Monitoring helps track remedial performance and the changes being wrought within the aquifer, so that ongoing remediation can be optimized.
7. Validate and close — Once the objective is achieved, stakeholders will usually require some degree of confirmation and certainty that the problems have been dealt with. Confirmation through validation sampling and monitoring and documentation that remediation has achieved the objectives are required. If appropriate, the site can be closed.

The mobility of contaminants in groundwater raises a number of issues for the setting of remedial objectives and assessing the most economic remediation approach alternative:

- Objectives must be framed in a temporal context — The level of risk associated with a given problem, and thus the predicted economic consequences should no action be taken, may change over time. In many cases, the longer we wait to deal with a problem, the worse it can get, and the more it may cost to deal with.
- Technology changes with time — What was considered technologically infeasible a decade ago may be wholly practicable and affordable today. This trend is bound to continue. In addition, the costs of remedial technologies may change with time.
- Regulations change with time — In the U.S., Europe, and the U.K., the regulations dealing with groundwater contamination have been evolving for the last several decades. Considering that planning horizons for serious groundwater contamination issues may be in the order of decades the likelihood is that relevant regulations and guidelines will change over the course of a project.
- Many deep groundwater contamination problems require long-term remedial solutions — In many cases, the only feasible remediation alternatives for groundwater contamination are containment and damage limitation, which involve long-term operation and maintenance (O&M) of remedial systems. Clearly, in these cases, time is a critical decision-making factor. Choosing an inappropriate planning horizon could compromise the decision-making process and result in selection of an infeasible and uneconomic remedial objective.

17.1.5 THE ECONOMIC VALUE OF GROUNDWATER

Groundwater has economic value because of:

- Its contribution to economic activities such as domestic, industrial, and commercial water use and irrigation for the production of crops or animal feed (direct use value).
- Its contribution to the hydrologic cycle (discharge to lakes, rivers, streams, wetlands, and other important surface water features) and through that recreation and amenity, (indirect use value).
- Its future uses by those who may or may not make use of groundwater at this time (option value).

These benefits are termed *use values* because they are related to how we make use of environmental resources like groundwater. People also value environmental resources irrespective of whether they make use of them now or plan to do so in future (termed *nonuse* values) so that:

- Other people can continue to use it (altruistic value).
- Future generations can have access to good-quality and sufficient ground-water (bequest value).
- The resource continues to exist in its own right irrespective of any human use made of it (existence value).

The sum of use and nonuse values is known as the total economic value (TEV).

Although the typology of the economic value of an environmental resource is relatively straightforward to establish, quantifying this value is not. This is mainly because the resources of concern are not traded in actual markets. They are external to the market mechanism. Therefore, there are no readily available data to estimate the full cost of using them or their TEV. When markets exist, as in the case of water, the price charged or may not cover all of the use values of the water and certainly does not cover the nonuse values. In fact, the market price that exists may even be distorted by subsidies, which are still used for water in many parts of the world.

Regardless of whether resources such as groundwater are traded in actual markets, their economic value of a resource is determined by what individuals are willing to pay to protect them from damage or to improve them (which is equivalent to buyers paying the prevailing price in actual markets) or what they are willing to accept in compensation to tolerate damage or forgo improvement (which is equivalent to sellers accepting a price in return for selling their products). What is common among the market price, *willingness to pay,* and *willingness to accept compensation* measures is that they are all in units of money. This common and familiar unit also enables us to compare the benefits of protecting a resource (the economic value of doing so) with the cost of this protection — or, in the case of groundwater, with protection and remediation of groundwater.

Economists have developed a number of economic valuation techniques that are used to quantify the total economic value. The first preference here is to use the *actual market price* (with caveats attached) because of the ease of accessing the data. Examples in the context of groundwater include the price of bottled water, public supply water, and so on. When actual market prices do not exist or are insufficient for the purposes of the analysis (e.g., they are distorted in some cases and exclude nonuse values in all cases), market price proxies could also be used. These are among the most popular techniques used in the literature about the economics of groundwater and include *avoidance costs* (i.e., the amount of money people spend to avoid the damage that is or may be caused by groundwater contamination). Two further techniques are revealed preference techniques and stated preference techniques. *Revealed preference techniques* investigate actual markets that do not trade the resource of concern but are influenced by it (e.g., house prices are influenced by the reliability and quality of water supply, as well as other structural, neighborhood, and environmental factors). *Stated preference techniques* create hypothetical markets by way of questionnaires through which individuals are given the chance to express their willingness to pay or willingness to accept compensation for the changes in the quality and quantity of the resource of concern.

There is a large and growing literature of economic valuation studies for many topics. Unfortunately, groundwater is not one of them. Most of the literature on groundwater is specific to site and contaminant and uses market proxies. Nevertheless, evidence from previous studies can be used in current remediation analysis if selected and adjusted carefully following the guidelines of the approach known as *benefits transfer*. This is especially the case for using the results of revealed and stated preference studies, because original data are easier to collect and more appropriate for using market price and market price proxies at the site and time of the economic analysis for remediation.

17.1.6 ECONOMIC ANALYSIS OF GROUNDWATER REMEDIATION

The main economic analysis technique recommended here is the cost–benefit analysis (CBA). CBA is a framework for comparing the monetary value of benefits of a project or policy with the monetary value of its costs. It can answer the two most important decision-making questions: "Should we remediate the contamination? If so, to what level of water quality?" The optimal remediation level, then, is the level at which the net benefit (benefits minus costs) of remediation is maximized.

In CBA, a *benefit* is defined as a change (financial, environmental, or social) that increases human well-being, and a *cost* as a change that decreases human well-being. The changes are measured against a common baseline (usually the do-nothing or business-as-usual scenario). In the context of groundwater remediation, benefits are the environmental damage avoided plus other benefits of clean-up. Costs, on the other hand, consist of financial and environmental costs of undertaking remediation.

Both costs and benefits occur over time (e.g., project lifetime), so for their comparison to be possible, costs and benefits that occur in different time periods must be expressed in relation to a given point in time. This point in time is the present, and the *present value* of costs and benefits is calculated using the discounting procedure. Discounting implies that the further into the future costs and benefits occur, the less valuable they are. How much less depends on the discount rate, about which some governments (including the U.K. and the U.S.) and international finance institutions issue guidance.

CBA compares present (or discounted) values of costs and benefits in two main ways. The first is the net present value (NPV) (benefits minus costs in present value terms); the second is the benefit–cost ratio (BCR) (benefits over costs in present value terms). If a remediation objective or approach has a positive NPV, economic analysis recommends its implementation. The objective or approach is also recommended if the BCR is greater than 1. Cost and benefit estimates should take risks and uncertainty into account economic and scientific.

CBA can be implemented from the point of view of the problem holder alone (financial analysis) and from the point of view of the whole society (economic analysis). The former is concerned only with those costs and benefits that affect the problem holder and reflect the preferences (including the discount rate) of the private party alone. The latter, on the other hand, is concerned with all costs and benefits (both internal and external) that affect the whole society (including the discount rate). At whatever level it is conducted, however, economic analysis is only one of the inputs to decision making and, hence options recommended by CBA, may or may not be implemented.

When an objective is predetermined and agreed upon by all stakeholders (e.g., imposed by legislation), there is usually no need to estimate benefits of reaching that objective. In other words, the first question of economic analysis — "Should anything be done at all?" — has already been answered. The exception is the cases in which different ways of achieving a given objective generate vastly different benefits for different stakeholders (or environmental assets). Because this is sometimes the case for remedial approaches, economic analysis is implemented for selecting the remedial approach and setting the remedial objective.

However, when we come to the selection of a remedial technology, we see that they do deliver the same or sufficiently similar benefits at different costs. In these cases, the appropriate economic approach is the cost-effectiveness analysis (CEA) or the least-cost analysis. CEA compares the capital, O&M, and external costs of remedial technologies and aims to identify the least-cost or the best-value-for-money option. In fact, groundwater remediation literature so far has used the term *cost–benefit analysis* to mean cost-effective analysis, where benefits refer to technical and other advantages of different remedial technologies.

17.1.7 REMEDIAL COSTS

The costs of implementing technical remedial solutions at specific sites where groundwater contamination exists are relatively well documented. Various sources in the literature provide information on the costs of implementing various remediation techniques. Consultancies, governments, and major corporations involved in managing and remediating contaminated sites have developed extensive databases on the costs of various remedial techniques for groundwater.

Most of the available groundwater literature dealing with the costs of remediation focuses on the problem holder's costs, known also as *private costs* of remediation. This is in part because private firms have developed considerable experience and knowledge of their costs and historically have had little impetus to focus on the wider issues.

Both the private (internal) and larger social (external) costs should be considered during remedial decision making and setting of remedial objectives. In situations in which the polluter has been identified as a private entity, the costs of implementing remediation will be borne wholly or substantially by that entity. However, society may also share some of the burden of cost of the remediation, should unmitigated effects to the wider environment occur as a direct result of the remediation. These are called the *external costs* of remediation.

Examples of some typical external costs of remediation include:

- Creating a new risk — In situations in which contaminants are removed from groundwater and introduced into another medium, a new risk, which did not previously exist, may result.
- Contamination of another medium — Certain remedial approaches may involve redirecting contamination to another medium, such as soil, air, or surface water.

- Contributing to air and greenhouse emissions — Any project that is energy intensive or that produces inordinate levels of greenhouse and other air emissions through the remedial process itself may also be producing external costs associated with climate change.
- Permanent elimination of water from the hydrologic cycle — If we assume that fresh water has some value, then a remedial process that removes it completely from the hydrologic cycle would produce a loss equivalent to the value of the volume of water processed.

External costs of remediation can be divided into two categories:

- Planned or process-related external costs that cannot or will not be mitigated against
- Unplanned, inadvertent, or unforeseen external costs

Planned external costs are increasingly being mitigated against. In many jurisdictions, specific regulatory measures are being put in place to ensure that remediation methods that deliberately shift costs from the problem holder to society are reduced or eliminated. But if the impact is an unplanned or unforeseen result of remediation, for which mitigation measures have not been provided or have not been successful in countering, the value of this damage is included as an *unplanned external cost* of remediation.

17.1.8 REMEDIAL BENEFITS

Benefits of remediation can also be seen in terms of private (internal) and public (external or wider economic) benefits. Private benefits include:

- Costs avoided if remediation takes place — These include avoiding the risk of litigation (and the considerable costs that may be involved), fines avoided, averting public relations damage (which could result in loss of sales revenue), and preventing control orders or shut-downs (which may result in lost production and revenue). The elimination of "stigma" value may also be relevant.
- Direct benefits — These might include increased property value or direct cost savings through access to clean groundwater.

These benefits can legitimately be included in a private (internal or financial) analysis of a remediation decision. However, not all of the private benefits are net increases in the wealth, well-being, or welfare of the society; some are simply transfers from one party in the society to another. For example, noncompliance fines avoided by the problem holder represent lost revenue for the competent authority imposing those fines.

Public or external economic benefits of groundwater remediation arise from the avoided damage to the environment and human health. The different ways in which remediation benefits accrue to the society, other than the problem holder, correspond to the components of the total economic value (i.e., use and nonuse values).

17.1.9 Using Cost–Benefit Analysis for Remedial Decision Making

The economics of groundwater remediation can be considered at four main levels:

- Policy objectives — Policy objectives are set by governments and are not the subject of this book. Therefore, decisions on what to remediate, what to protect, and what to sacrifice must be generally guided by the policy of the day. Policy could include maximization of human welfare, for instance. In many jurisdictions, including the U.K., Europe, Canada, and the U.S., stated national environmental policy is based on the protection of human health.
- Remedial objective — Setting the remedial objective (or risk-management objective) for a given contamination problem should be based on the results of risk assessment. Only a limited number of remedial objectives are available: the receptor is protected, impacts to the receptor are reduced or eliminated, the contamination is removed, or contamination is reduced to a set, predetermined regulatory level. The remedial objective is the level at which the benefits of remediation are most readily and fundamentally determined. Benefits are tied to the fundamental objective and the approach used to achieve it.
- Remedial approach — The remedial approach focuses on ways to break the risk linkage that causes damage. Remedial approaches remove the source; eliminate the pathway; or protect, move, or manage the receptor. Because different approaches are likely to lead to different types and levels of benefits and have different costs, CBA is usually required (rather than the least-cost or cost-effectiveness analysis).The analysis of remedial approaches provides a link between remedial objectives and the hundreds of remedial technologies available. Also, the degree to which the linkage is broken, the timing of the action, and the spatial location at which the action is taken are all variables that must be considered when choosing the approach. A constraints analysis can be undertaken to help assess which approaches can realistically be achieved.
- Remedial technology — The remedial technology selection level involves choosing the most cost-effective way of putting a remedial approach into play to achieve a remedial objective. External costs of remediation should also be incorporated into the cost analysis.

A structured framework has been developed for using CBA to choose the most economically eficient remedial objectives for a given problem. In practice, only the most high-profile and difficult sites will warrant use of the whole framework. Some steps are more demanding and more difficult than others, and some require information that is not always available. More commonly, certain steps can be skipped, and the analysis can focus on only the critical and readily executed steps. The framework allows a gradual screening of objective alternatives. Initial screening is based on the results of risk assessment, application of basic policy, and constraint criteria. At this stage, a short-list of practical remedial objectives (or a clear objective) is selected. Suitable remedial approaches are identified that can reach each objective, and the benefits of each are assessed. Then, the least-cost way to implement each approach is determined. This leads to conducting a high-level CBA or partial CBA (through inclusion of nonmonetary criteria) to compare the costs and benefits of each remedial approach for each objective. Then, remedial objectives can be compared in terms of their net benefits. Alternatively, the net benefit of each remedial approach can be considered, and the objectives met by each approach evaluated. Either way, what results is a remedial approach, or combination of approaches, that manages the risks identified in the most economic way possible.

17.2 CONCLUSIONS

17.2.1 LEVEL OF EFFORT

For large, complex problems, full analysis using all of the framework steps involves a considerable amount of effort. In circumstances involving complex and serious risks, large expected expenditures, and high public profile, such effort may well be required and advisable. However, for the majority of smaller, more straightforward sites, the process will be much more manageable and far less intensive.

17.2.2 INTERDEPENDENCY OF SITE REMEDIATION AND GROUNDWATER REMEDIATION

Groundwater contamination problems cannot be considered in isolation. Contaminated land and soil issues at a site must also be considered, because these are often the sources of additional contaminants migrating to groundwater through the unsaturated zone. Therefore, groundwater remediation CBA cannot be conducted in isolation and yield meaningful results. For instance, the effect of source remediation by soil treatment at a site may have significant positive effects on groundwater quality, which cannot be ignored. The costs and benefits of source removal on property value cannot readily be uncoupled from those of groundwater remediation, nor should they be. Setting realistic and economically sound remedial objectives for groundwater is likely, in most instances, to involve the whole site.

17.2.3 EFFECT OF DISCOUNT RATE

Using lower discount rates in CBA will, all other factors remaining equal, result in higher benefits from resource protection or remediation but not necessarily in higher net benefits, because the present value of costs would also be higher — more so in circumstances in which remedial activities require significant ongoing expenditure. Remedial approaches that involve forward-weighted expenditure, however, will tend to realize benefits sooner and incur costs sooner. For these approaches, net benefits will tend to increase with declining discount rate, all other factors remaining equal.

Problem holders with a commercial perspective will be more likely to defer expenditure on remediation, all other factors remaining equal, when their present value costs are high and their (private) benefits are low. On the other hand, society can take a longer-term view and value future costs and benefits more than problem holders can. This is one reason why economic analysis conducted from the point of view of society uses a lower discount rate. The main discount rate used in the analysis should follow the prevailing guidance at the time of the analysis and also should be tested through sensitivity analysis to show the effect of the different views of different stakeholders.

17.2.4 EFFECT OF DEFERRAL

Deferring remediation may result in a drop in net benefits for some remedial approaches and in increases for others, depending on the relative time flows of costs and benefits. Net benefits will tend to decline as a result of deferral for approaches in which expenditure and benefit are one-time, fixed-sum events, closely spaced. In general, this will apply for approaches with intensive forward-weighted expenditure and more immediate realization of benefits. The same occurs for approaches where benefit flows over time exceed cost flows or where benefit flows are time-delayed with respect to expenditures. In these cases, benefits suffer proportionately more than costs as a result of deferral.

For remedial approaches in which the cost flows over time exceed benefit flows over time, and if remedial benefits and costs accrue preferentially over the longer term, net benefit will tend to *increase* if remediation is deferred. In situations in which damage is increasing with time (as with a mobile and expanding plume), delays in remediation are much more likely to mean a reduction in net benefit.

17.2.5 FURTHER RESEARCH ON THE ECONOMIC VALUE OF AQUIFERS

Most of the current economic valuation literature on water-related issues concentrates on surface water and its recreational uses, whereas most of the literature on groundwater is specific to site and contaminant and uses market proxies. Further research in the field should present all aspects of groundwater, the services it provides, and the threats it is under to capture the full spectrum of total economic value.

According to the available evidence, TEV of groundwater is greater than the abstraction or user fees paid for it. With increasing scarcity of fresh groundwater that is relatively cheap to abstract, it is likely that this difference between the two measures will increase. If economic analysis of remediation continues to be limited to market price data alone, it will be increasingly likely that benefits of remediation are underestimated and environmental damage that should be stopped will be allowed to continue, to the detriment of the environment and society.

17.2.6 USE AND LIMITATIONS

Following the framework methodology presented in this book allows users to consider more fully the relationships between various possible remedial objectives and the risks they seek to manage. The interdependence of many of these objectives and the widely varying economic costs and benefits these will produce show the importance of conducting some form of CBA.

Limitations to economic analysis in this context will include uncertainty over the future, commodity and resource prices, people's preferences for environmental quality, changes in remedial technology and regulatory requirements, and the need for further research into the economic value of the aquifer.

Although one alternative is to refrain from undertaking any cost–benefit analysis unless the preceding uncertainties are elemenated, we believe that, when carefully and transparently estimated, a number for external costs and benefits included in the decision-making process is better than no number. The process of undertaking economic analysis allows for the wider definitions of economic costs and benefits to be taken into account when making remedial decisions, even if the actual numbers are not used. This, on its own, can be considered a success.

Glossary of Economic Terms

Altruistic value Person A's willingness to pay for the continued enjoyment of person B's use of environmental resources.

Avertive expenditures (or avoidance cost) Expenditures undertaken to avoid or mitigate the impacts of pollution.

Consumer surplus The difference between the amount paid for a good or service and the maximum amount that an individual would be willing to pay.

Contingent valuation A survey technique used to derive values for environmental change by estimating individuals' willingness to pay (or to accept compensation) for a specified change in the quality and quantity of a resource.

Cost–benefit analysis A form of economic analysis in which costs and benefits over time, expressed in monetary units, are compared.

Discounting Converts costs and benefits occurring at different points in future into comparable units of today (present value).

Existence values Values that result from an individual's desire to ensure that an environmental asset is preserved for its own sake (a type of nonuse value).

Externalities Changes that are not reflected in actual market prices; uncompensated impacts that affect third parties. Goods that remain unpriced and thus are external to the market (i.e., free goods such as those relating to the environment, with an example being clean air).

Financial analysis Aimed at determining the financial gains and losses due to a policy or a project.

Hedonic pricing method: An implicit price for an environmental attribute is estimated from consideration of the actual markets, which are influenced by the quality and quantity of the environmental resource of concern (e.g., water quality improvements and property values).

Market price approach In a perfectly competitive market, the market price of a good provides an appropriate estimate of its economic value (excluding nonuse values). In markets that are not perfectly competitive, economic value is calculated by removal of subsidies or other price distortions.

Net present value The present value (i.e., in year 0) of the difference between the discounted stream of benefits and the discounted stream of costs.

Nonuse value A value that is not related to direct or indirect use of the environment (e.g., existence, altruistic, and bequest values).

Opportunity cost The value of a resource in its next best alternative use.

Option value Value to a consumer of retaining the option to consume a good or sevice in the future.

Replacement costs Impacts on environmental assets are measured in terms of the cost of replacing or re-creating the asset.

Total economic value The sum of use values (direct use, indirect use, and option) plus nonuse values (altruistic, bequest, and existence).

Transfer payment A payment for which no good or service is obtained in return (e.g., a tax or subsidy).

Travel cost method The benefits arising from the recreational use of a site (or cost of collecting natural resources like firewood and water) are estimated in terms of the costs incurred in travel to the site.

Uncertainty Stems from a lack of information or scientific knowledge and is characteristic of all predictive assessment.

Use value A value related to the actual direct or indirect use of the environment (e.g., recreational value).

Willingness to accept compensation (WTA) The amount of money an individual would be willing to accept as compensation for forgoing a benefit or tolerating a cost.

Willingness to pay (WTP) The amount of money an individual would be willing to pay to secure a benefit or avoid a cost.

Index

A

Access to water in developing countries, 4
Accidents/fatalities and landfilling wastes
        excavated from contaminated sites,
        150
Actual market value technique, 69–70, 80–81, 316
Advection process and groundwater
        contamination, 39
Africa, 3–6
Agriculture as a contributor to groundwater
        pollution, 5–6, 16, 29
Allocation policies, water, 6
Altruistic value, 67
American Petroleum Institute, 16
Ammonium, 263
Apportionment matrix, benefits, 189, 239
Approaches, remedial; *see also* Benefits,
        remedial; Cost listings;
        Technologies, remedial
    gas plant site in the U.K.
        available remedial approaches, 215, 216
        cost function, 235, 236
        economically optimal approach, 235, 237
        objectives, relationships among multiple,
          219–224
        overview, 214–215
        short-list, developing a, 215–222
    MtBE-contaminated aquifer in the U.S.,
        266–267
    objectives, reaching remedial, 181–185,
        193–195
    other than economic tools, 109–112
    overview, 170–171
    source-pathway-receptor linkage, 113–114
    tritium-contaminated groundwater, 272–275
Aquifers, 13, 38, 120–122, 322; *see also*
        Contaminated listings; MtBE-
        contaminated aquifer in the U.S.;
        individual subject headings
Aral Sea, 6
Aromatics, 42–43
Assiniboine Delta aquifer, 127–128
ASTM, 53–54
Australia, 6
Avertive expenditures, 70
Avoidance costs, 70, 81, 127, 129–132, 316

B

Balance in the hydrologic cycle, 35
Basal containment layer, accidental piercing of,
        153–155
Baseflow, 35
Bedrock, 33–35
Bench-scale experiments, 56
Benefits, remedial; *see also* benefits *under*
        Objectives, reaching remedial and
        Tritium-contaminated groundwater;
        Cost-benefit analysis; Economic
        theory for groundwater remediation
    apportionment matrix, 189
    benefit-cost ratio, 100
    defining terms, 78
    education, 163
    financial vs. economic analysis, 106
    gas plant site in the U.K.
        apportionment matrix, 239
        approach/objective, possible benefit of
          each, 225, 227–231
        nonquantifiable benefits, apply, 235–236
        nonquantifiable benefits, identify/assess,
          233–234
        quantifiable benefits, identify/assess,
          231–232
        risk assessment, 225, 226
        threshold benefits, 232–233
    literature, benefits term different in technical
        and economic, 141–142
    MtBE-contaminated aquifer in the U.S.,
        265–266
    overview, 157
    private (internal), 8, 157–159, 227–228, 277
    problem and solution, example, 301–303
    property value increase, 161–163
    public (external), 159–161, 229–230, 277
    summary/conclusions, 319–320
    time, over, 162–163
    transfer, benefits, 73, 76–77, 317
    tritium-contaminated groundwater, 291
Benzene, 15, 258–262, 264
Bequest value, 67
Biological systems used to treat contaminants, 58
Bioremediation, 58
Bioventing, 58